HELLO

So it's now time to get started... this workbook/revi_ _ _ _ _ _ _ _ _ _
the crossover content and concentrates on stuff that is Higher tier only.
But first, a few housekeeping rules about how to use this guide:

1 Make sure you have the correct equipment. You will need the usual maths stuff and we also recommend you use a notebook / paper alongside this workbook to make notes and for more working out if needed.

2 Remind yourself of the key skills on the pages titled 'Key Skills'. These pages can help you refresh your memory of the crossover topics, so identify any that you need to revise.

3 Look through the command word glossary on the 'Command Word' page. These words are the key words that can appear in a question and will instruct you on how to tackle the problem.

4 Work through the 'Ready' section of a double page, making notes if needed. Then have a go at the 'Set' and 'Go' questions (no cheating with the answers at the back... they're for when you've finished).

5 Return to the same double page a couple of days later and check you still remember the content. Then and only then are you allowed to tick the checklist and rip off the corner of the page.*

if it's not your workbook... check with the owner!

Also included... a set of cut-out flashcards to help you remember the stuff you need to know (you can always make more)

On each double-page there are three sections for you to work through:

READY? Read the key information and work carefully through the examples. Sometimes highlighting is used to give extra guidance and look out for the maths police who point out common misconceptions!

SET? Dive right in and have a go at these questions. They are closely linked to the worked examples (no curveballs yet). Check your answers.

GO! Here we go... some more questions, but this time exam-style (curveballs included). Again, don't forget to check your answers.

You CAN do this... now let's get started!

You're almost ready to get started ... but first you must decide where you will work:

Curled up under a duvet ❌

In front of the TV ❌

At a well-lit desk with a proper chair ✔

Turn to page ...		READY?	SET?	GO!
Estimating with Powers & Roots	12			
Recurring Decimals	14			
Fractional Indices	16			
Product Rule for Counting	18			
Calculating with Bounds	20			
Surds 1	22			
Surds 2	24			
Algebraic Fractions 1	26			
Solving Harder Equations	28			
Factorising	30			
Expanding Brackets	32			
Algebraic Fractions 2	34			
Algebraic Proof	36			
Functions 1	38			
Functions 2	40			

Turn to page ...		READY?	SET?	GO!
Perpendicular Lines	42			
Completing the Square	44			
Quadratic Functions	46			
Quadratic Equations 1	48			
Quadratic Equations 2	50			
Exponential Graphs	52			
Trigonometric Graphs	54			
Transforming Graphs 1	56			
Transforming Graphs 2	58			
Estimating Gradients	60			
Rates of Change	62			
Area Under a Graph	64			
Simultaneous Equations 1	66			
Simultaneous Equations 2	68			
Rearranging Formulae	70			

What does success look like?

Tick the checklist when you've completed each section

Check a few days later ... still confident?

Tear off the corner thingy

Ace!

And now for a couple of non-sticky notes about this book:
1) Look out for the non-calculator symbol ... no cheating!
2) Diagrams are only drawn to scale when it says so.
3) Some questions refer to students' workings, shown in blue font.

Contents and Checklist

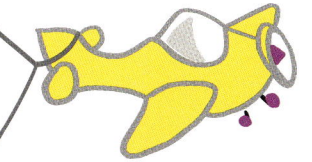

Always check your answers ... page 135 onwards

Know the formulae you need to remember (cut out the flashcards)

Check you know what the command words mean

Make sure you have all the correct equipment

You can do this!

	Turn to page ...	READY?	SET?	GO!
Iteration	72			
Inequalities 1	74			
Inequalities 2	76			
Sequences 1	78			
Sequences 2	80			
Direct Proportion	82			
Inverse Proportion	84			
Enlargements	86			
Transformations	88			
Circle Theorems 1	90			
Circle Theorems 2	92			
Circle Theorems 3	94			
Equation of a Circle 1	96			
Equation of a Circle 2	98			
Similarity	100			

	Turn to page ...	READY?	SET?	GO!
Solids	102			
Trigonometry in 3D	104			
The Sine Rule	106			
The Cosine Rule	108			
Area of a Triangle	110			
Geometric Proof	112			
Vectors 1	114			
Vectors 2	116			
Probability 1	118			
Probability 2	120			
Statistics	122			
Box Plots	124			
Cumulative Frequency	126			
Histograms	128			
Capture-Recapture	130			

It's no good just owning this book ...

use it!

3

KEY SKILLS

Number properties

Multiples are the result of multiplying a number by an integer, e.g. multiples of 6 are 6, 12, 18, 24, ...

Factors divide into another number without a remainder, e.g. factors of 12 are 1, 2, 3, 4, 6 and 12

Integer: A positive or negative whole number, or zero, e.g. -3, 0 or 5 etc.

Prime numbers: A number with exactly two factors 2, 3, 5, 7, 11, 13, 17 etc.

Square number: The result of multiplying a whole number by itself e.g. $3 \times 3 = 9$

Cube number: The result of multiplying a whole number by itself, then by itself again e.g. $4 \times 4 \times 4 = 64$

The **highest common factor** (HCF) of two or more numbers is the largest factor that they have in common

The **lowest common multiple** (LCM) of two or more numbers is the smallest multiple that they have in common

Need to revise ☐ Nailed it! ☐

Fractions

Addition
When the denominators are not equal, we need to find equivalent fractions to help
e.g. to work out $\frac{3}{7} + \frac{2}{5}$
$$\frac{3}{7} + \frac{2}{5} = \frac{15}{35} + \frac{14}{35} = \frac{29}{35}$$

Subtraction
The process is the same as when adding
e.g. $\frac{12}{13} - \frac{3}{4} = \frac{48}{52} - \frac{39}{52} = \frac{9}{52}$

Multiplication
Fractions can be multiplied using the fact: $\frac{a}{b} \times \frac{c}{d} = \frac{a \times c}{b \times d}$
e.g. $\frac{4}{5} \times \frac{7}{9} = \frac{4 \times 7}{5 \times 9} = \frac{28}{45}$

Division
Dividing by a proper fraction is the same as multiplying by the **reciprocal** of the fraction
e.g. $\frac{5}{7} \div \frac{3}{4} = \frac{5}{7} \times \frac{4}{3} = \frac{20}{21}$

Treat any **integers** as fractions with a denominator of 1, e.g. $9 = \frac{9}{1}$

Convert any **mixed numbers** into improper fractions
e.g. $3\frac{2}{5} = 3 + \frac{2}{5} = \frac{15}{5} + \frac{2}{5} = \frac{17}{5}$

Need to revise ☐ Nailed it! ☐

Symbols

$\equiv$ Identical to
$\approx$ Approximately equal to
$>$ Greater than
$\geq$ Greater than OR equal to
$=$ Equal to
$<$ Less than
$\neq$ Not equal to
$\leq$ Less than OR equal to

Need to revise ☐ Nailed it! ☐

Fraction of an amount

Finding $\frac{1}{3}$ of 18
$\frac{1}{3} = 1$ of 3 equal parts
The whole is 18
so $\frac{1}{3}$ of $18 = 18 \div 3 = 6$

Finding $\frac{2}{5}$ of 40
The value of one part $= 40 \div 5 = 8$
$\frac{1}{5}$ of $40 = 8$ so $\frac{2}{5}$ of $40 = 2 \times 8 = 16$

Need to revise ☐ Nailed it! ☐

NUMBER

Know your calculator

$\sqrt{\square}$ finds any root of a number

Ans uses the previous answer in the next question

Powers are either $x^\square$ or $\wedge$

Use ▣ or ab/c for mixed numbers

Convert between types of number with S⇔D or FORMAT

Need to revise ☐ Nailed it! ☐

Rounding

First significant figure ↘ Second significant figure ↙
13.578
First decimal place ↗ Second decimal place ↖

e.g. 13.578 to 1 d.p. = 13.6   13.578 to 2 s.f. = 14
(Note: the first significant figure in 0.00**4**71 is 4)

The **error interval** for a rounded number uses the lower and upper bounds: e.g.
If a number x rounds to 3.7 to 1 d.p. then
$$3.65 \leq x < 3.75$$
Lower bound   Upper bound

Need to revise ☐ Nailed it! ☐

Standard form

A number greater than or equal to 1, but less than 10
$$9.73 \times 10^4$$
Always multiplied by 10 with a positive or negative power

For example:
$9.73 \times 10^4 \equiv 97\,300$
$1.66 \times 10^{-5} \equiv 0.000\,016\,6$

Need to revise ☐ Nailed it! ☐

Linear sequences

A **linear sequence** increases (or decreases) in constant steps

1, 7, 13, 19, 25
  +6 +6 +6 +6

The nth **term** of a linear sequence is connected to this constant step

The nth term of 6, 12, 18, 24, 30 is $6n$

The nth term of 1, 7, 13, 19, 25 is $6n - 5$ $\}-5$

Need to revise ☐ Nailed it! ☐

4

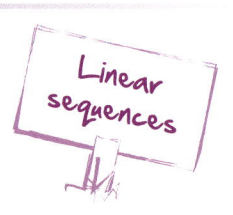

$n + 2$ means 'n' add 2
$2n$ means 2 times 'n'
$n - 2$ means 'n' subtract 2
$\frac{n}{2}$ means 'n' divided by 2
$2 - n$ means 2 subtract 'n'
n^2 means 'n' squared

The language of algebra

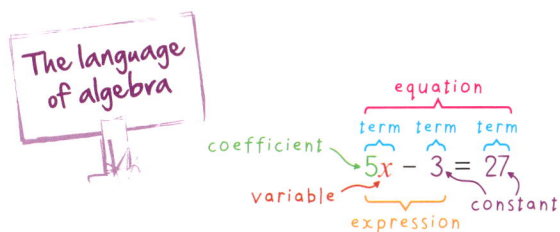

In algebra, a **constant** is a number on its own, e.g. 3 is a constant in $5x - 3$

A **variable** is a quantity that varies in value. They are usually represented by a letter, often x or y.

A **coefficient** is the number in front of a variable (multiplying it), e.g. 5 is the coefficient of $5x$

A **term** can be a constant, a variable, or either of these things multiplied together, e.g. 5, x, $5x$

An **expression** is a mathematical statement made by adding and/or subtracting terms, e.g. $5x - 3$

A mathematical statement containing an equals symbol is called an **equation**, e.g. $5x - 3 = 27$

A **formula** is a type of equation that shows the relationship between different quantities, e.g. $C = \pi d$

An equation that is always true is called an **identity**, e.g. $2x + 3x = 5x$, which should really be written as:
$$2x + 3x \equiv 5x$$

Need to revise ☐   Nailed it! ☐

KEY SKILLS

1st Law of indices: $a^m \times a^n = a^{m+n}$ (add powers)
2nd Law of indices: $a^m \div a^n = a^{m-n}$ (subtract powers)
3rd Law of indices: $(a^m)^n = a^{m \times n}$ (multiply powers)
4th Law of indices: $a^{-m} = \dfrac{1}{a^m}$

Index laws

Need to revise ☐   Nailed it! ☐

 Don't forget... anything to the power of zero equals 1
$5^0 = 1$ $567^0 = 1$ $y^0 = 1$ etc.

ALGEBRA

To **multiply out** $4(2x + 5)$

×	$2x$	5
4	$8x$	20

$4 \times 2x$ 4×5

So $4(2x + 5) = 8x + 20$

Expanding brackets

To **expand and simplify** $(x + 5)(x - 3)$

×	x	5
x	x^2	$5x$
-3	$-3x$	-15

$-3x + 5x = 2x$

So $(x + 5)(x - 3) = x^2 + 2x - 15$

Need to revise ☐   Nailed it! ☐

TECHNICAL STUFF: $(x + 5)(x - 3)$ is equivalent to $(x + 5)(x + -3)$

Factorising is the process of finding factors. It is the opposite of expanding:

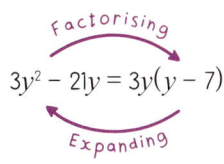

Factorising → $3y^2 - 21y = 3y(y - 7)$ ← Expanding

Factorising expressions

To fully factorise $18x + 24y$

×	$3x$	$4y$
6	$18x$	$24y$

$6 \times 3x$ $6 \times 4y$

So $18x + 24y = 6(3x + 4y)$

To factorise $n^2 - 2n - 35$

×	n	5
n	n^2	$5n$
-7	$-7n$	-35

$-7n + 5n = -2n$

So $n^2 - 2n - 35 = (n - 7)(n + 5)$

Need to revise ☐   Nailed it! ☐

Every straight line can be written as $y = mx + c$, where m is the gradient and c is the y-intercept

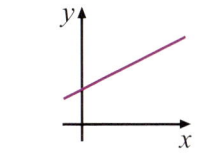

A **positive** gradient goes upwards from left to right

Straight line graphs

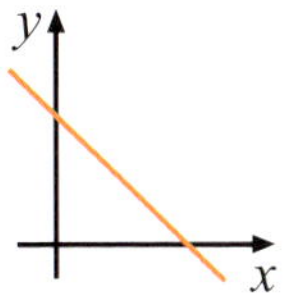

A **negative** gradient goes downwards from left to right

Need to revise ☐   Nailed it! ☐

Drawing inequalities

Need to revise ☐   Nailed it! ☐

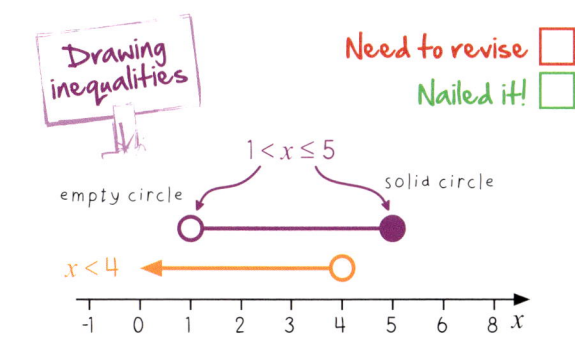

$1 < x \leq 5$
empty circle   solid circle
$x < 4$

Quadratic graphs have the general shape:

Non linear graphs

These curves could appear stretched or squeezed

Cubic graphs have the general shape:

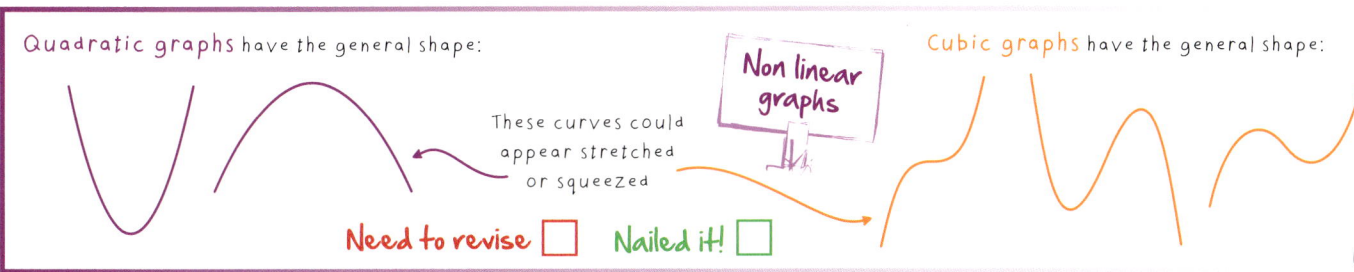

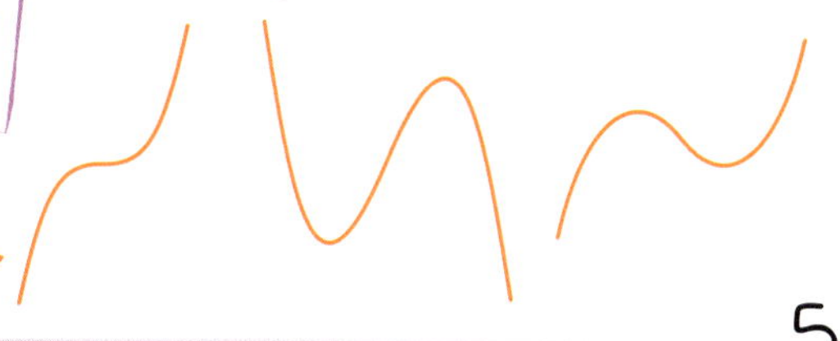

Need to revise ☐   Nailed it! ☐

5

Area

The area of a **rectangle** can be found using the formula:

Area of rectangle = base × height

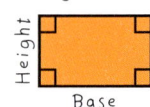

This can be used to find formulae for other shapes:

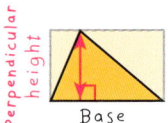

Area of **triangle** = $\frac{1}{2}$ × base × perpendicular height

Area of **parallelogram** = base × perpendicular height

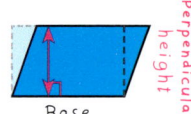

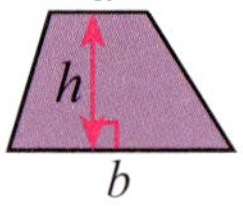

Area of **trapezium** = $\frac{1}{2} \times (a+b) \times h$ or $\frac{a+b}{2} \times h$

Need to revise ☐
Nailed it! ☐

GEOMETRY AND MEASURES

Constructions

The **perpendicular bisector** of the line segment AB:

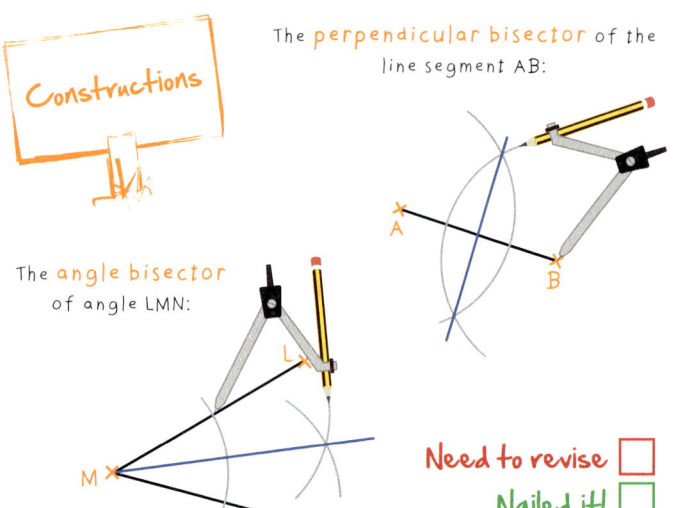

The **angle bisector** of angle LMN:

Need to revise ☐
Nailed it! ☐

Angle and line facts

Isosceles triangles: two equal sides and two equal angles

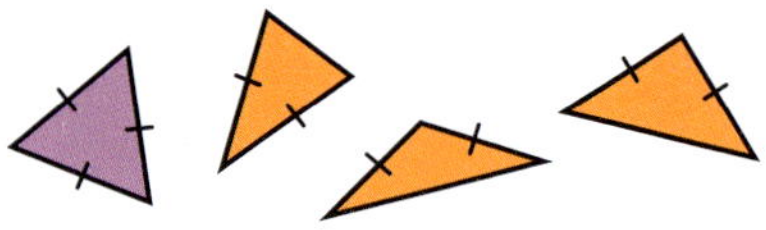

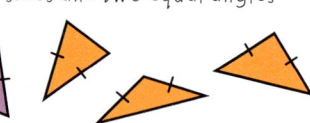

Equilateral triangles: all sides equal and all angles 60°

Angles in a triangle add up to 180°

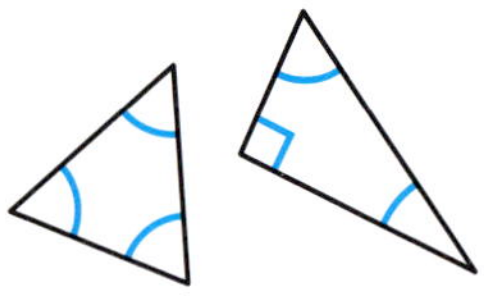

Angles in a quadrilateral add up to 360°

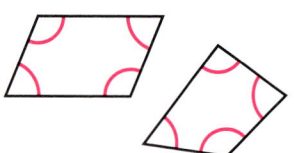

Angles meeting at a point add up to 360°

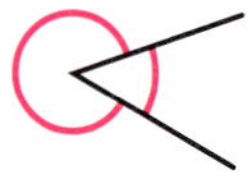

Angles meeting at a point on a straight line add up to 180°

Vertically opposite angles are equal

Co-interior angles add up to 180°

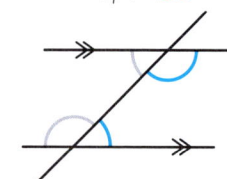

Alternate angles are equal

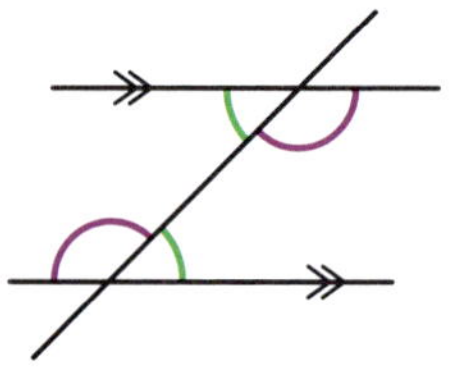

Corresponding angles are equal

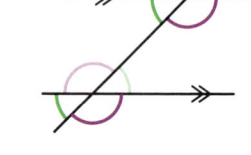

Need to revise ☐   Nailed it! ☐

3D shapes

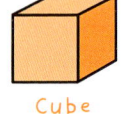 **Cube**
Volume = length³

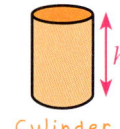 **Cylinder**
Volume = $\pi r^2 h$

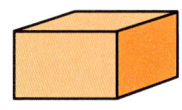 **Cuboid**
Volume = length × width × height

Key
r = radius
h = height
l = slant height

Triangular prism
Volume = area of cross-section × length

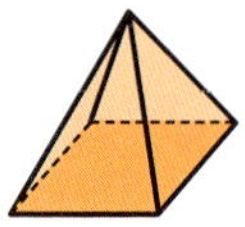

 **Square-based pyramid**
Volume = $\frac{1}{3}$ × base area × height

 **Sphere**
* Volume = $\frac{4}{3}\pi r^3$
* Surface area = $4\pi r^2$

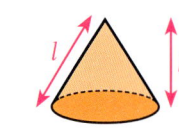 **Cone**
* Volume = $\frac{1}{3}\pi r^2 h$
* Curved surface area = $\pi r l$

* Some formulae may be given in the exam

Need to revise ☐
Nailed it! ☐

Polygons

The total of a polygon's **exterior angles** is always 360°

The sum of the **interior angles** can be found by splitting into triangles

Sides	Name	Angle sum
5	Pentagon	540°
6	Hexagon	720°
7	Heptagon	900°
8	Octagon	1080°
9	Nonagon	1260°
10	Decagon	1440°

(+180° between each)

Need to revise ☐   Nailed it! ☐

KEY SKILLS

Circles
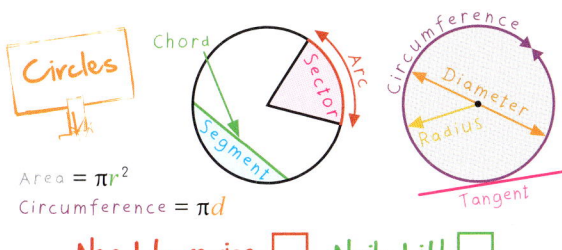
Area = πr^2
Circumference = πd

Need to revise ☐   Nailed it! ☐

Pythagoras' theorem
$c^2 = a^2 + b^2$
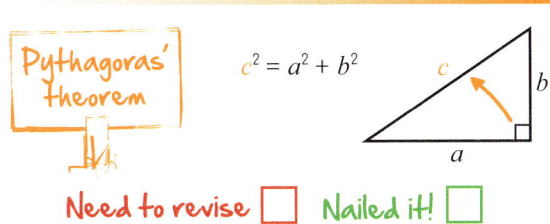

Need to revise ☐   Nailed it! ☐

Congruence
Congruent shapes are identical. The corresponding side lengths are equal and the corresponding angles are also equal.
Two triangles are congruent if:

Three sides are equal.
Side, Side, Side → SSS

Two angles AND a corresponding side are equal.
Angle, Angle, Side → AAS

Both have a right angle, each hypotenuse is the same length AND another side is also equal.
Right-angle, Hypotenuse, Side → RHS

Two sides are equal AND the angle between those sides is also equal.
Side, Angle, Side → SAS

Need to revise ☐   Nailed it! ☐

Describing transformations

Rotation: Centre of rotation, angle and direction
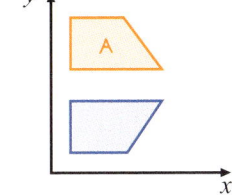
Reflection: Equation of the mirror line

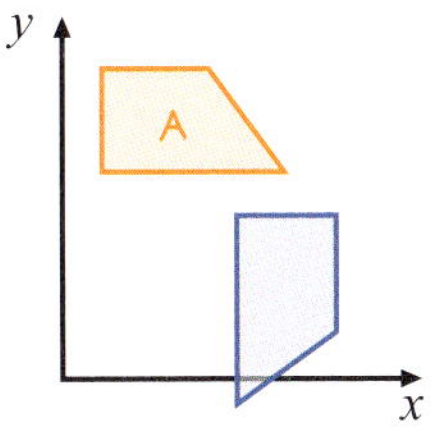

Translation: Vector (distance left/right and up/down)
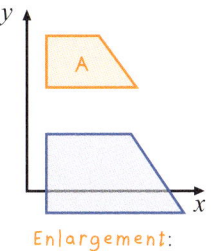
Enlargement: Centre of enlargement and scale factor

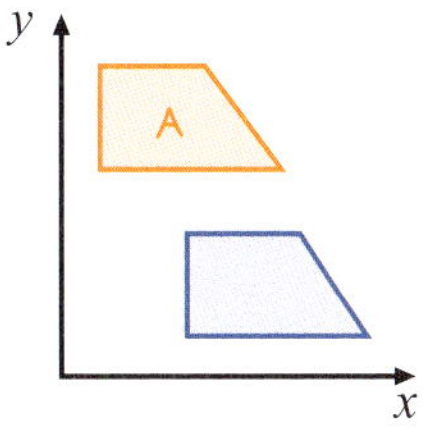

Need to revise ☐   Nailed it! ☐

Trigonometry
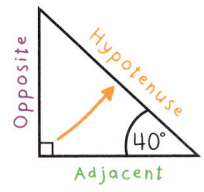   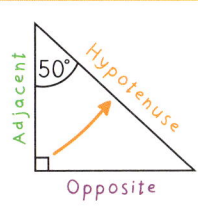

Right-angled triangles can be labelled according to a chosen angle (not the right angle)

The trigonometric functions sine (sin), cosine (cos) and tangent (tan) connect the angle and side lengths:

chosen angle → $\sin\theta = \dfrac{O}{H}$ $\cos\theta = \dfrac{A}{H}$ $\tan\theta = \dfrac{O}{A}$

Need to revise ☐   Nailed it! ☐

Bearings
Bearings are:
- Measured clockwise
- Measured from NORTH
- Given as three digits

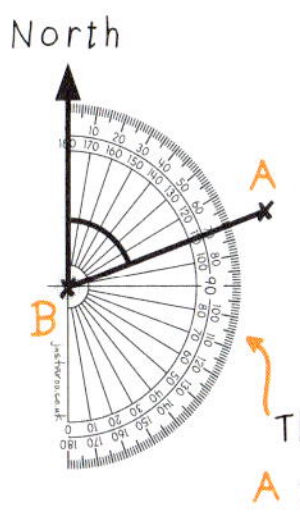

The bearing of A from B is 070°

Need to revise ☐   Nailed it! ☐

Trigonometry: exact values

θ	0°	30°	45°	60°	90°
$\sin\theta$	0	$\dfrac{1}{2}$	$\dfrac{\sqrt{2}}{2}$	$\dfrac{\sqrt{3}}{2}$	1
$\cos\theta$	1	$\dfrac{\sqrt{3}}{2}$	$\dfrac{\sqrt{2}}{2}$	$\dfrac{1}{2}$	0
$\tan\theta$	0	$\dfrac{\sqrt{3}}{3}$	1	$\sqrt{3}$	Doesn't exist

Need to revise ☐   Nailed it! ☐

Ratio
Mel and Chris share some money in the ratio 5 : 3

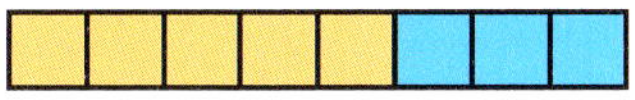

★ If the total is £40, how much does Chris get? 
| 5 | 5 | 5 | 5 | 5 | 5 | 5 | 5 | 40 ÷ (5 + 3)

★ If Mel gets £40, how much does Chris get? 
| 8 | 8 | 8 | 8 | 8 | 8 | 8 | 8 | 40 ÷ 5

★ If the difference in amounts is £40, how much does Chris get? 
| 20 | 20 | 20 | 20 | 20 | 20 | 20 | 20 | 40 ÷ (5 − 3)

Need to revise ☐   Nailed it! ☐

Metric units
10 mm = 1 cm
100 cm = 1 m
1000 m = 1 km
1000 ml = 1 litre
1000 cm³ = 1 litre
1000 g = 1 kg

Need to revise ☐   Nailed it! ☐

Compound measures

$\text{speed} = \dfrac{\text{distance}}{\text{time}}$ $\text{density} = \dfrac{\text{mass}}{\text{volume}}$

Need to revise ☐   Nailed it! ☐

$\text{population density} = \dfrac{\text{number of people}}{\text{area}}$ $\text{pressure} = \dfrac{\text{force}}{\text{area}}$

7

KEY SKILLS

Percentages

For **percentage change** problems:

$$\text{Percentage change} = \frac{\text{actual change}}{\text{original amount}} \times 100$$

For **growth or decay** problems:

$$\text{Total} = \text{starting number} \times (\text{multiplier})^{\text{repeats}}$$

For growth: multiplier > 1
For decay: multiplier < 1

Need to revise ☐ Nailed it! ☐

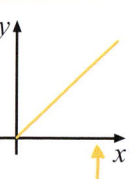
Proportion

Two variables are **directly proportional** if they increase (or decrease) at the same rate as each other.

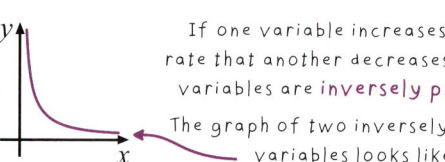

The graph of two directly proportional variables looks like this.

If one variable increases at the same rate that another decreases, then the two variables are **inversely proportional**.

The graph of two inversely proportional variables looks like this.

Need to revise ☐ Nailed it! ☐

RATIO AND PROPORTION

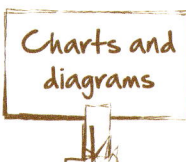
Charts and diagrams

```
1 | 4 5 8
2 | 1 3 6 9 9
3 | 0 5 7
4 | 4
```

Key: $2 \mid 1 = 21$ kg

A **stem and leaf diagram** is a way of representing and ordering data. Each value is split into a 'stem' and a 'leaf'. Stem and leaf diagrams MUST include a key.

In a **frequency tree**, each 'node' (circle) is equal to the sum of the other circles that branch off from it.

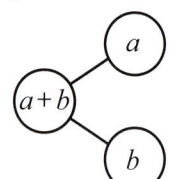

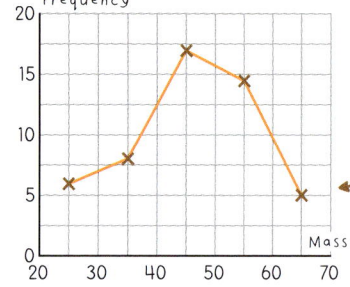

A **frequency polygon**, shows the frequencies plotted against the midpoint of each group.

Need to revise ☐ Nailed it! ☐

Averages

Mode → most common
Median → middle, when arranged in order
Mean → the total shared between the number of items

When data is presented in a table, the subtotals are needed to work out the mean:

Goals	Frequency	Subtotals
0	× 6	0
1	× 8	8
2	× 17	34
3	× 14	42
4	× 5	20
TOTALS:	50	104

$$\text{Mean} = \frac{104}{50} = 2.08$$

When data is grouped, we use the midpoints to estimate the actual values:

Mass (m kg)	Frequency	Midpoint	Subtotals
$20 \leq m < 30$	6	× 25	150
$30 \leq m < 40$	8	× 35	280
$40 \leq m < 50$	17	× 45	765
$50 \leq m < 60$	14	× 55	770
$60 \leq m < 70$	5	× 65	325
TOTALS:	50		2290

$$\text{Estimate for the mean} = \frac{2290}{50} = 45.8 \text{ kg}$$

Need to revise ☐ Nailed it! ☐

Probability

If all outcomes are **equally likely**:

$$\text{Probability} = \frac{\text{number of required outcomes}}{\text{total number of outcomes}}$$

Need to revise ☐ Nailed it! ☐

STATISTICS AND PROBABILITY

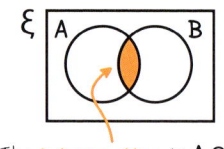
The **intersection** is $A \cap B$

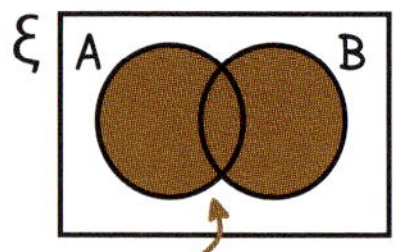

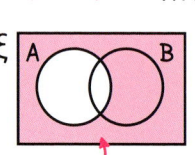
The **union** is $A \cup B$

The **complement** of A is A'
(everything not in set A)

Venn diagrams

Need to revise ☐ Nailed it! ☐

Types of **correlation**: the closer the points are to the line of best fit, the stronger the correlation.

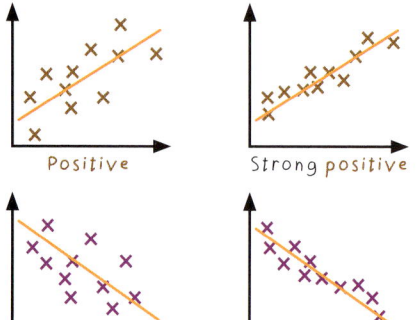

Positive   Strong positive

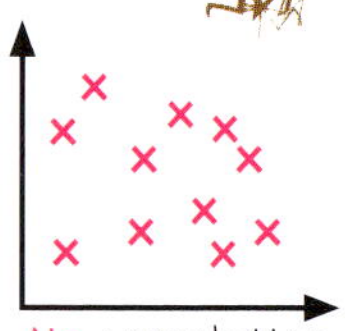

Negative   Strong negative

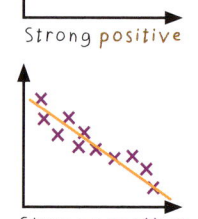
Scatter diagrams

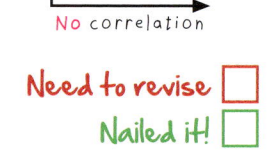
No correlation

Need to revise ☐ Nailed it! ☐

Flashcards to cut out

The equation of a circle	Multiplication with surds
The formula for solving quadratic equations	Division with surds
Volume of a pyramid	Fractional indices
Volume of a cone	The sine rule
The angle at the centre ...	The cosine rule
Angles in the same segment ...	Area of a triangle
The angle in a semicircle ...	Graph of $y = \sin x$
Two tangents from an external point ...	Graph of $y = \cos x$
A radius and a tangent that meet ...	Graph of $y = \tan x$
A radius that is perpendicular to a chord ...	The effect of $y = f(x) + a$ on the graph of $y = f(x)$
The alternate segment tells us that ...	The effect of $y = f(x + a)$ on the graph of $y = f(x)$
The opposite angles in a cyclic quadrilateral ...	The effect of $y = f(-x)$ on the graph of $y = f(x)$
Box plots	The effect of $y = -f(x)$ on the graph of $y = f(x)$

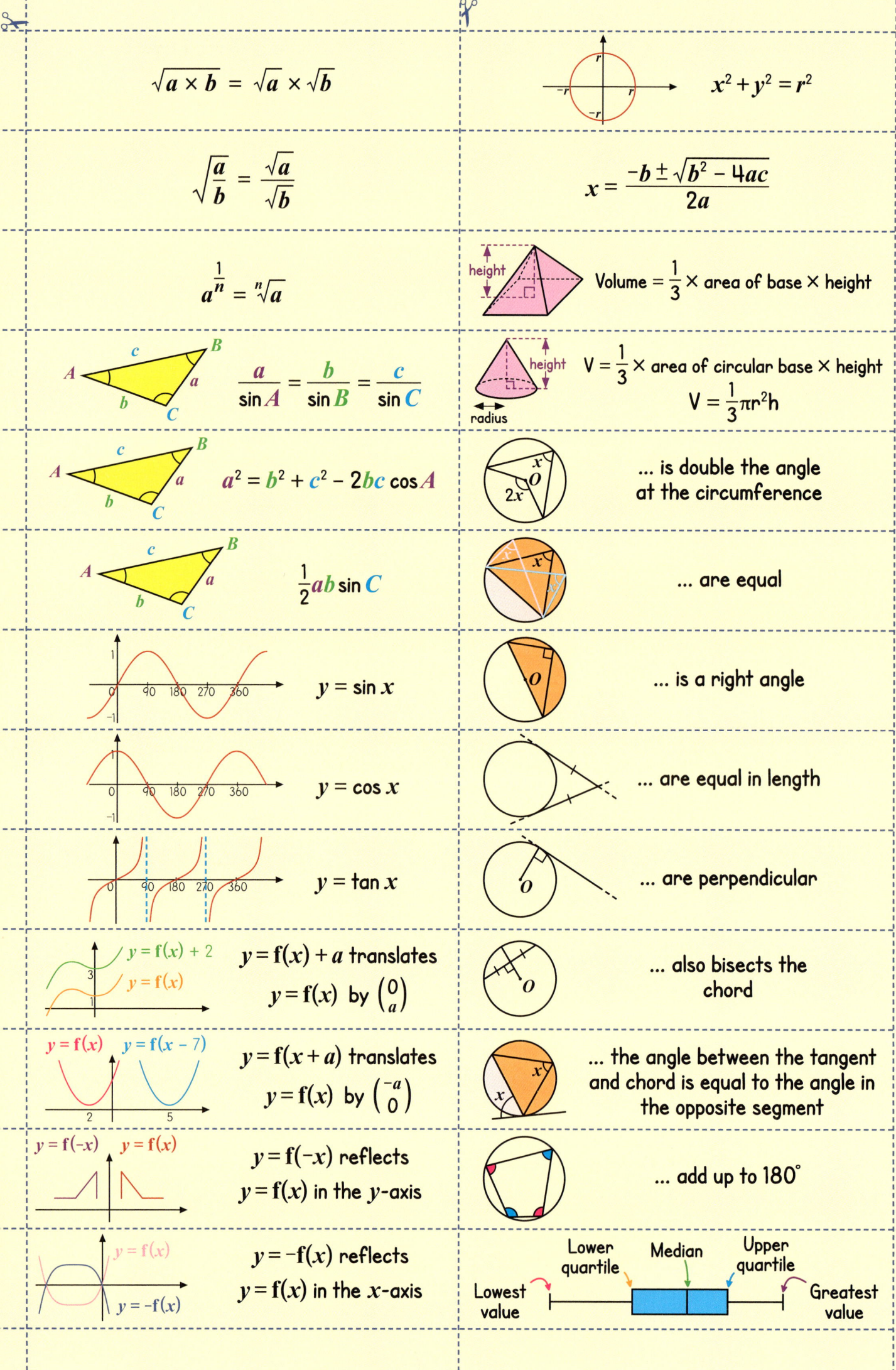

COMMAND WORDS

EXPAND — Remove brackets from an algebraic expression
$5(2x + 3) = 10x + 15$

In maths, 'expand' does NOT mean E x p a n d

EXPRESS — Rewrite in another form
Express $2^3 \times 2^5$ as a power of 2: 2^8

EVALUATE — Find the VALUE
Evaluate 4^3: $4 \times 4 \times 4 = 64$

EXPLAIN — Give reasons to support the decision or the answer

WRITE — Give the answer without needing to show working out
Write $\frac{3}{4}$ as a decimal: 0.75

SIMPLIFY — Make an algebraic expression simpler by collecting like terms:
$3x + 4 + 2x = 5x + 4$
Make a fraction simpler by cancelling common factors:
$\frac{12}{16} = \frac{3}{4}$

SOLVE — Find the answer to a problem
Solve: $2x + 13 = 35$, $x = 11$

FACTORISE — Put brackets into an algebraic expression
$x^2 + 6x + 8 = (x + 2)(x + 4)$
$15y + 12 = 3(5y + 4)$

ROUND (GIVE YOUR ANSWER TO) — Make a number simpler but keep its value close to what it was
74.26 rounded to …
2 significant figures is 74
1 decimal place is 74.3

CALCULATE or WORK OUT — Perform one or more steps to get an answer
Calculate 15% of £40:
10% → £4 so 5% → £2
15% → £6

Calculate does NOT always mean we have to use a calculator

ORDER — Use a rule to arrange
Order from smallest to largest:
$\sqrt{45}$, $\sqrt[3]{20}$, 4.33
$\sqrt[3]{20}$, 4.33, $\sqrt{45}$

ESTIMATE — Give a sensible approximate answer using rounding
Estimate 21.7×6.3:
$20 \times 6 = 120$

CONSTRUCT — Create an accurate drawing using the correct maths equipment

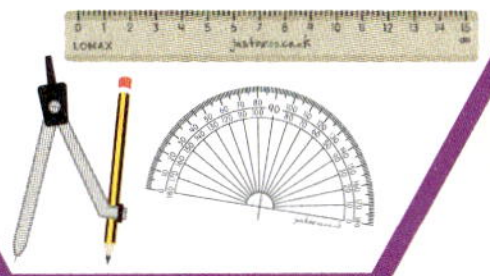

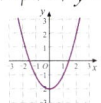

DRAW — Create an accurate drawing of a shape or diagram

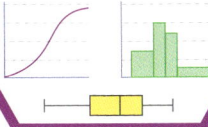

RATIONALISE — Make a number rational by removing surds
Rationalise the denominator of $\frac{4}{\sqrt{7}}$:
$\frac{4}{\sqrt{7}} = \frac{4 \times \sqrt{7}}{\sqrt{7} \times \sqrt{7}} = \frac{4\sqrt{7}}{7}$

SKETCH — Create a rough drawing that shows key features
Sketch the graph of $y = 2^x$

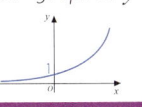

No need to use a ruler or a pair of compasses

SHOW THAT — Use reasons or logic to explain why a given fact or statement is true
e.g. Show that the equation $x^3 + 6x - 1 = 0$ has a solution between $x = 0$ and $x = 1$

PLOT — Mark a point or draw a graph on a set of axes
Plot the graph of $y = x^2 - 2$:

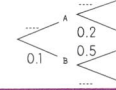

DESCRIBE — Use correct maths vocabulary to explain key features
Rotation 180° centre (0,0)

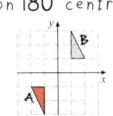

FIND — Work out an answer to a problem
Find the mode of 6, 3, 9, 5, 3: Mode = 3

COMPLETE — Fill in missing values in a table such as
$y = 2x + 1$
… and on a diagram such as

GIVE REASONS or JUSTIFY — Use reasons to explain thinking, such as
the angles on a straight line add up to 180°

REPRESENT — Display information in a chart or graph, such as a scatter graph

COMPARE — Describe the similarities and differences between two (or more) things

PROVE — Create a convincing argument using a logical chain of steps

READY?

We can combine our knowledge of powers and place value to square or cube multiples of 10. For example:

$40^2 = 40 \times 40 = 4 \times 10 \times 4 \times 10 = 16 \times 100 = 1600$

$200^3 = 200 \times 200 \times 200$
$= 2 \times 100 \times 2 \times 100 \times 2 \times 100$
$= 8 \times 1\,000\,000 = 8\,000\,000$

Sometimes we can find roots of multiples of 10. It is helpful to look for square numbers and cube numbers as clues. For example:

$\sqrt{3600}$ — 36 is a square number

$6 \times 6 = 36$ and $60 \times 60 = 3600$ so $\sqrt{3600} = 60$

$\sqrt[3]{8000}$ — 8 is a cube number

$2 \times 2 \times 2 = 8$ and $20 \times 20 \times 20 = 8000$
so $\sqrt[3]{8000} = 20$

TECHNICAL STUFF: $\sqrt{}$ means 'the positive square root of'

e.g. 1 Estimate the value of $\sqrt{28}$ to one decimal place.

1. Find two consecutive square numbers that are either side of 28:
1, 4, 9, 16, **25, 36**, 49, ...

2. Use the square root of these numbers to make a first estimate:
$\sqrt{25} = 5$ and $\sqrt{36} = 6$ so $\sqrt{28}$ is between 5 and 6

3. Use the position of 28 in relation to 25 and 36 to make a more accurate estimate:
28 is closer to 25 than 36, but not very close so say $\sqrt{28} \approx 5.3$

≈ means 'approximately equal to' (between 5.2 and 5.4 would be fine)

CHECK-IN

1. Put a ring around the cube numbers:
3   8   27   36   64   81   1000

2. Work out:
$\sqrt{81}=$ $\sqrt{0.49}=$
$\sqrt[3]{1\,000\,000}=$ $\sqrt[4]{81}=$

3. Write down all the square numbers between 100 and 200

e.g. 2 Estimate the value of $\sqrt[3]{60}$ to two significant figures.

1. Find two consecutive cube numbers that are either side of 60:
1, 8, **27, 64**, 125, ...

2. Use the cube root of these numbers to make a first estimate:
$\sqrt[3]{27} = 3$ and $\sqrt[3]{64} = 4$ so $\sqrt[3]{60}$ is between 3 and 4

3. Use the position of 60 in relation to 27 and 64 to make a more accurate estimate:
60 is very close to 64 so say $\sqrt[3]{60} \approx 3.9$

We can also use estimation skills to work with powers:

e.g. 3 Choose the best estimate for the value of 42^3

126   1600   42\,000   64\,000   125\,000

42 = 40 to one significant figure
$40^3 = 40 \times 40 \times 40 = 4 \times 4 \times 4 \times 1000 = 64\,000$

e.g. 4 Use estimation to help choose the correct value of 738^2

478\,904   544\,644   528\,860   695\,312

$700^2 = 490\,000$ and $800^2 = 640\,000$
so 478\,904 is too small and 695\,312 is too big

The last digits are also a useful clue...
738×738 must have a last digit of 4 since $8 \times 8 = 64$
So 738^2 must equal 544\,644

A. Estimate the value to one decimal place:
(i) $\sqrt{41}$
(ii) $\sqrt{23}$
(iii) $\sqrt{70}$
(iv) $\sqrt{105}$
(v) $\sqrt[3]{20}$
(vi) $\sqrt[3]{68}$

B. Choose the best estimate for the value of 58^3
18\,000   125\,000   166\,000   216\,000

C. Choose the best estimate for the value of 208^2
40\,000   62\,500   80\,000   90\,000

D. Use estimation to help choose the correct value of 45^3
64\,000   91\,125   125\,000   130\,125

E. Use estimation to help choose the correct value of 34^4
628\,846   1\,336\,336   1\,628\,864   2\,418\,806

F. Use < or > to complete the statements:
(i) $\sqrt{14}$ ◯ $\sqrt[3]{80}$
(ii) $\sqrt[3]{57}$ ◯ $\sqrt{32}$
(iii) $\sqrt{45}$ ◯ $\sqrt[3]{123}$
(iv) $\sqrt{0.08}$ ◯ $\sqrt[3]{0.08}$

GO! ESTIMATING WITH POWERS AND ROOTS

Terrific!

1. Madge is asked to estimate the value of
   $$\sqrt[3]{0.066}$$
   She writes:
   $$\sqrt[3]{0.066} = 0.022$$
   Explain why Madge is not correct.

2. The Cube is a 24-storey building in the centre of Birmingham. The main section of the building is a perfect cube with side lengths of 53.1 metres.
   a) Estimate the volume of this section of the building.
   b) Is your answer an overestimate or an underestimate? Explain your answer.

3. The infield in a game of baseball is a square.

   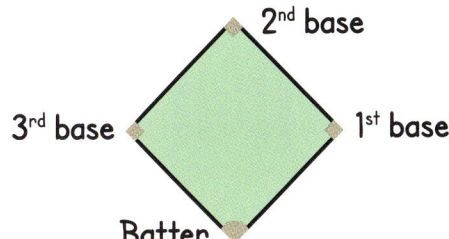

   The area of the square is 752 m². Estimate the perimeter of the square.

4. In electronics, a formula connecting power in watts (P), voltage in volts (V) and resistance in ohms (R) is:
   $$P = \frac{V^2}{R}$$
   Estimate the value of P if:
   $V = 130$ volts
   $R = 20.9$ ohms

Some fractions can be written as terminating decimals. For example:

$\frac{7}{25} = 0.28$ (denominator: 5^2)

$\frac{29}{40} = 0.725$ (denominator: $2^3 \times 5$)

$\frac{643}{400} = 1.6075$ (denominator: $2^4 \times 5^2$)

If 2s and 5s are the only prime factors of the denominator, then the decimal is terminating

Lots of fractions are equivalent to recurring decimals. We use dot notation* to show these recurring decimals:

$\frac{1}{3} = 0.3333... = 0.\dot{3}$ $\frac{8}{11} = 0.7272... = 0.\dot{7}\dot{2}$

*sometimes a bar is used instead; e.g. $\frac{1}{3} = 0.\overline{3}$

We can write any fraction as a decimal using division.

e.g. $\frac{5}{12} \rightarrow 12 \overline{)5.^50^20^80^80^80...} = 0.41666...$ so $\frac{5}{12} = 0.41\dot{6}$

It is possible to convert a recurring decimal into a fraction (using algebra to help).

e.g. 1 Write $0.\dot{3}\dot{6}$ as a fraction in its simplest form.

Remember: $0.\dot{3}\dot{6} = 0.363636...$

1 Represent the decimal with a letter: Let $x = 0.363636...$ *(The decimal parts match)*

2 Multiply by 10 to make some related facts:
Then $10x = 3.636363...$
and $100x = 36.363636...$ *

3 Choose two facts* and subtract to eliminate the recurring decimal:

$100x - x = 36.363636... - 0.363636...$

$99x = 36$ the decimal part has been eliminated!

4 Solve the equation: $x = \frac{36}{99} = \frac{4}{11}$ so $0.\dot{3}\dot{6} = \frac{4}{11}$

CHECK-IN

1. Write using recurring decimal notation:
   $0.5555555... =$ $0.43434343... =$
   $0.0171717... =$ $0.754754754... =$

2. Simplify the fractions:
   $\frac{24}{90}$ $\frac{117}{999}$ $\frac{408}{990}$

Ready for take off?

Sometimes we need to multiply x by 1000 too:

e.g. 2 Express $0.6\dot{1}\dot{8}$ as a fraction in its simplest form.

1 Represent: Let $x = 0.6\dot{1}\dot{8}$ $0.6181818...$

2 Make related facts:
Then $10x = 6.\dot{1}\dot{8}$ * $6.181818...$
$100x = 61.8\dot{1}$ $61.818181...$
and $1000x = 618.\dot{1}\dot{8}$ * $618.181818...$

3 Choose* and eliminate:

$1000x - 10x = 618.\dot{1}\dot{8} - 6.\dot{1}\dot{8}$

$990x = 612$

4 Solve: $x = \frac{612}{990} = \frac{306}{495} = \frac{102}{165} = \frac{34}{55}$ so $0.6\dot{1}\dot{8} = \frac{34}{55}$

(÷2, ÷3, ÷3)

Some recurring decimals are greater than 1 ...

e.g. 3 Prove algebraically that $1.\dot{4}6\dot{2} = \frac{487}{333}$

1 Let $x = 0.\dot{4}6\dot{2}$ * $0.462462...$

2 Then $10x = 4.6\dot{2}\dot{4}$ $4.624624...$
$100x = 46.2\dot{4}\dot{6}$ $46.246246...$
and $1000x = 462.\dot{4}6\dot{2}$ * $462.462462...$

3 $1000x - x = 462.\dot{4}6\dot{2} - 0.\dot{4}6\dot{2}$

$999x = 462$

4 $x = \frac{462}{999} = \frac{154}{333}$ so $1.\dot{4}6\dot{2} = 1\frac{154}{333} = \frac{487}{333}$

Deal with the decimal part first

A. Tick the fractions that are equivalent to recurring decimals:

$\frac{5}{9}$ $\frac{13}{80}$ $\frac{9}{14}$ $\frac{4}{65}$

B. Express as decimals:

(i) $\frac{3}{11}$

(ii) $\frac{7}{12}$

(iii) $\frac{5}{18}$

(iv) $\frac{23}{30}$

C. Write the decimals as fractions in their simplest form:

(i) $0.\dot{3}\dot{8}$

(ii) $0.\dot{5}\dot{7}$

(iii) $0.\dot{2}\dot{6}$

D. Prove algebraically that:

(i) $0.\dot{4}1\dot{7} = \frac{139}{333}$

(ii) $0.9\dot{2}\dot{7} = \frac{51}{55}$

(iii) $0.34\dot{6} = \frac{26}{75}$

SET?

E. Express the decimals as fractions in their simplest form:

(i) $1.\dot{4}$

(ii) $2.9\dot{6}$

(iii) $3.0\dot{3}\dot{9}$

14

RECURRINGINGINGINGING DECIMALS

Lovely!

1. Imogen writes:

   $0.5\dot{4} = \frac{54}{99} = \frac{6}{11}$

   Do you agree with Imogen? Explain why.

2. A batsman's strike rate in cricket is calculated as the number of runs scored per 100 balls.

   The New Zealand player, Luke Ronchi, once scored 170 runs off 99 balls against Sri Lanka.

   What was Ronchi's strike rate in this game as a recurring decimal?

   HINT: Find the number of runs per ball first

3. Find the value of each missing number:

   a) $4.7\dot{2} = \frac{85}{a}$

   b) $6.\dot{3}\dot{9} = \frac{b}{33}$

   c) $c - 0.\dot{4} = 2.\dot{1}\dot{5}$

4. Calculate $0.\dot{7} \times 1.4\dot{3}$

   Give your answer as a mixed number in its simplest form.

READY?

When working with indices (also called powers), there are some rules that we should already know:

1st Law of indices: $a^m \times a^n = a^{m+n}$ (add powers)

2nd Law of indices: $a^m \div a^n = a^{m-n}$ (subtract powers)

3rd Law of indices: $(a^m)^n = a^{m \times n}$ (multiply powers)

4th Law of indices: $a^{-m} = \dfrac{1}{a^m}$

Don't forget... anything to the power of *zero* equals 1
$5^0 = 1$ $567^0 = 1$ $y^0 = 1$ etc.

We can use the first law to work out other facts. For example, because 8 is equal to a number multiplied by itself and itself again ($2 \times 2 \times 2$):

$$8 = 8^1 = 8^{\frac{1}{3}+\frac{1}{3}+\frac{1}{3}} = 8^{\frac{1}{3}} \times 8^{\frac{1}{3}} \times 8^{\frac{1}{3}}$$

Also: $8 = 2 \times 2 \times 2 = \sqrt[3]{8} \times \sqrt[3]{8} \times \sqrt[3]{8}$

So it must be true that: $\sqrt[3]{8} = 8^{\frac{1}{3}}$

25 is equal to a number multiplied by itself (5×5), so:

$$25 = 25^1 = 25^{\frac{1}{2}+\frac{1}{2}} = 25^{\frac{1}{2}} \times 25^{\frac{1}{2}}$$

Also: $25 = 5 \times 5 = \sqrt{25} \times \sqrt{25}$

So it is true that: $\sqrt{25} = 25^{\frac{1}{2}}$

We could write $\sqrt{25}$ as $\sqrt[2]{25}$

In general, the **5th Law of indices** is $a^{\frac{1}{n}} = \sqrt[n]{a}$

e.g. 1 Write down the value of $144^{\frac{1}{2}}$

$144^{\frac{1}{2}} = \sqrt[2]{144} = 12$ *Invisible 2*

CHECK-IN

1. Write as a single power of 9:
$9^4 \times 9^3 =$ $9^4 \div 9^3 =$
$(9^5)^3 =$ $\dfrac{9^6 \times 9}{9^6} =$

2. Write down the value of 2^{-3}:

3. Simplify:
$(3x^5 y^2)^2 =$ $\dfrac{24a^8 b^6}{8a^5 b} =$

Ready for take off?

---

e.g. 2 Work out $16^{\frac{1}{4}}$

$16^{\frac{1}{4}} = \sqrt[4]{16} = 2$ $2 \times 2 \times 2 \times 2 = 16$

e.g. 3 Evaluate $1000^{\frac{1}{3}}$

$1000^{\frac{1}{3}} = \sqrt[3]{1000} = 10$

We can combine the laws of indices to solve problems:

e.g. 4 Evaluate $8^{\frac{2}{3}}$

1. Use the third law of indices to break up the $\frac{2}{3}$

$\frac{2}{3} = \frac{1}{3} \times 2$

$8^{\frac{2}{3}} = \left(8^{\frac{1}{3}}\right)^2$

We could also use $8^{\frac{2}{3}} = \left(8^2\right)^{\frac{1}{3}}$

2. Evaluate: $\left(8^{\frac{1}{3}}\right)^2 = \left(\sqrt[3]{8}\right)^2 = 2^2 = 4$

e.g. 5 Evaluate $4^{-\frac{5}{2}}$

1. Use the third law of indices to break up the $-\frac{5}{2}$

$-\frac{5}{2} = \frac{1}{2} \times 5 \times -1$

$4^{-\frac{5}{2}} = \left(\left(4^{\frac{1}{2}}\right)^5\right)^{-1}$

Other orders could be used

2. Evaluate: $\left(\left(4^{\frac{1}{2}}\right)^5\right)^{-1} = \left(\left(\sqrt{4}\right)^5\right)^{-1} = \left(2^5\right)^{-1} = 32^{-1} = \dfrac{1}{32}$

Sometimes a fraction is raised to a fractional power...

e.g. 6 Work out the value of $\left(\dfrac{8}{125}\right)^{-\frac{2}{3}}$

$-\frac{2}{3} = \frac{1}{3} \times 2 \times -1$

1. Break up the $-\frac{2}{3}$

$\left(\dfrac{8}{125}\right)^{-\frac{2}{3}} = \left(\left(\left(\dfrac{8}{125}\right)^{\frac{1}{3}}\right)^2\right)^{-1}$

$\frac{2}{5} \times \frac{2}{5} \times \frac{2}{5} = \frac{8}{125}$

2. $\left(\left(\left(\dfrac{8}{125}\right)^{\frac{1}{3}}\right)^2\right)^{-1} = \left(\left(\sqrt[3]{\dfrac{8}{125}}\right)^2\right)^{-1} = \left(\left(\dfrac{2}{5}\right)^2\right)^{-1} = \left(\dfrac{4}{25}\right)^{-1} = \dfrac{25}{4}$

SET?

A. Evaluate:

(i) $81^{\frac{1}{2}}$

(ii) $27^{\frac{1}{3}}$

(iii) $64^{\frac{1}{3}}$

(iv) $81^{\frac{1}{4}}$

(v) $\left(\dfrac{1}{1000}\right)^{\frac{1}{3}}$

(vi) $\left(\dfrac{4}{9}\right)^{\frac{1}{2}}$

B. Work out the value of:

(i) $64^{\frac{2}{3}}$

(ii) $16^{\frac{3}{4}}$

(iii) $32^{\frac{2}{5}}$

(iv) $25^{\frac{3}{2}}$

(v) $\left(\dfrac{125}{8}\right)^{\frac{2}{3}}$

(vi) $\left(\dfrac{27}{1000}\right)^{\frac{4}{3}}$

C. Evaluate:

(i) $125^{-\frac{2}{3}}$

(ii) $4^{-\frac{5}{2}}$

(iii) $10\,000^{-\frac{3}{4}}$

(iv) $\left(\dfrac{9}{25}\right)^{-\frac{3}{2}}$

16

GO! FRACTIONAL INDICES

sterling!

1. Nick is asked to work out the value of $4^{\frac{1}{2}}$

   He writes:

   $4^{\frac{1}{2}} = 4 \times \frac{1}{2} = 2$

   Explain the mistake that Nick has made.

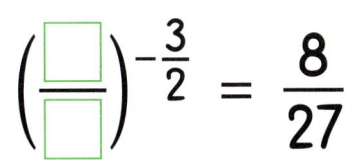

2. Write digits in each box to make the statement correct.

   $\left(\dfrac{\Box}{\Box}\right)^{-\frac{3}{2}} = \dfrac{8}{27}$

   HINT: Rewrite 8 and 27 as powers

3. Find the value of the missing numbers:

   a) $\sqrt[3]{a} = 3^{\frac{4}{3}}$

   b) $\sqrt[4]{8} = b^{\frac{3}{4}}$

   c) $5 \times \sqrt{125} = 25^{c}$

4. Write $\sqrt[3]{4} \times \sqrt{8}$ as a single power of 2

   $\sqrt[3]{4} \times \sqrt{8} = 2^{\Box}$

Sometimes we can solve problems by using multiplication as a shortcut to counting. For example:

A pencil case contains red, blue, black and green pens. It also contains pink, blue, yellow and purple highlighters. One pen and one highlighter are chosen.

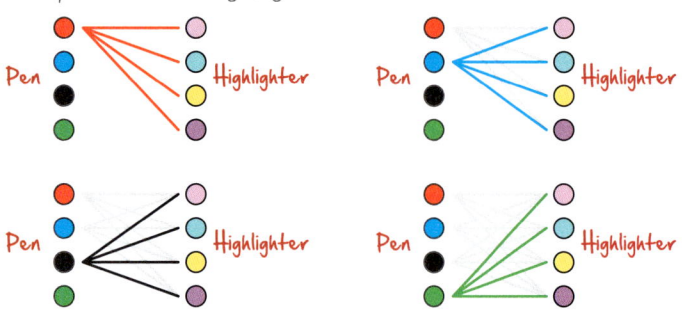

There are 4 options for the pen. Each of these 4 options can be paired with each of the 4 options for the highlighter.

So there must be 4 × 4 = 16 combinations in total.

This process is useful when the numbers are large. It is often known as the **product rule for counting**.

e.g. 1 A class contains 14 boys and 16 girls. One boy and one girl are chosen to be members of a school council. How many combinations of people can be chosen?

There are 14 different boys that could be chosen — 14 × ...

Each of the 16 girls could be paired with each of the 14 boys: — 14 × 16

14 × 16 = 224

224 possible combinations of people can be chosen

Remember: product means multiply

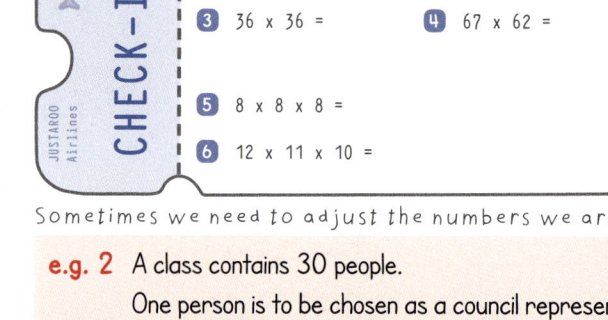

1	40 × 40 =	2	20 × 19 =
3	36 × 36 =	4	67 × 62 =
5	8 × 8 × 8 =		
6	12 × 11 × 10 =		

Sometimes we need to adjust the numbers we are given:

e.g. 2 A class contains 30 people.
One person is to be chosen as a council representative and a different person to be a sports representative. How many different ways are there of choosing people for these roles?

There are 30 different ways to choose the council representative — 30 × ...

For each of these options there are 29 people remaining who could be the sports representative — 30 × 29

30 × 29 = 870

There are 870 ways of choosing people for these roles

At other times we need to adjust the product by thinking about the context of the problem.

e.g. 3 16 football teams compete in a tournament. Each team plays every one of the other teams once. How many matches are played in total?

Each of the 16 teams plays every one of the other 15 teams — 16 × 15

But ... this counts every match twice; e.g. Team A v Team B is the same match as Team B v Team A.

We need to adjust by dividing by 2.

$$\frac{16 \times 15}{2} = \frac{240}{2} = 120$$

120 matches are played in total

A. There are 78 boys and 75 girls in a year group of a school.
A head boy and a head girl are to be elected.
How many different combinations of people can be chosen?

B. A cricket team is made up of 11 players.
One player is chosen as the captain and another player as the vice-captain.
How many ways are there of choosing players for these roles?

C. A combination lock uses a code with three digits.
Each digit can be any of the numbers from 0 to 9.
How many different codes can be chosen?

D. A large box contains chocolates wrapped in foil.
The foil is either red, black, white, gold, blue, green or pink.
How many ways are there to choose two chocolates with different colour foil?

E. There are 52 cards in a full pack. All the cards are different.
Adam takes 3 of the cards.
How different ways of choosing these cards are there?

GO! PRODUCT RULE FOR COUNTING

Astounding!

1. One boy and one girl from Year 11 are chosen to represent their school.

   There are 58 boys in Year 11.

   There are 3886 possible combinations of boys and girls.

   How many girls are in Year 11?

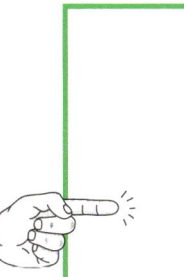

2. A code is made as follows:

   | Letter | Digit | Digit | Digit | Letter |

   - Each digit can be any number from 0 to 9
   - Each letter can be any of the 26 letters in the alphabet.

   Neil works out the number of possible codes as follows:

   $$26 \times 10 \times 9 \times 8 \times 25 = 468\,000$$

   Neil is not correct. Explain why.

3. The vehicle registration plate system in the UK is based on three pieces of information:

   Age identifier

   MW12 OCB

   Area code   Random letters

   - There are 434 different area codes.
   - None of the random letters are I or Q.
   - The age identifier changes every 6 months.

   How many different registration plates are possible every 6 months?

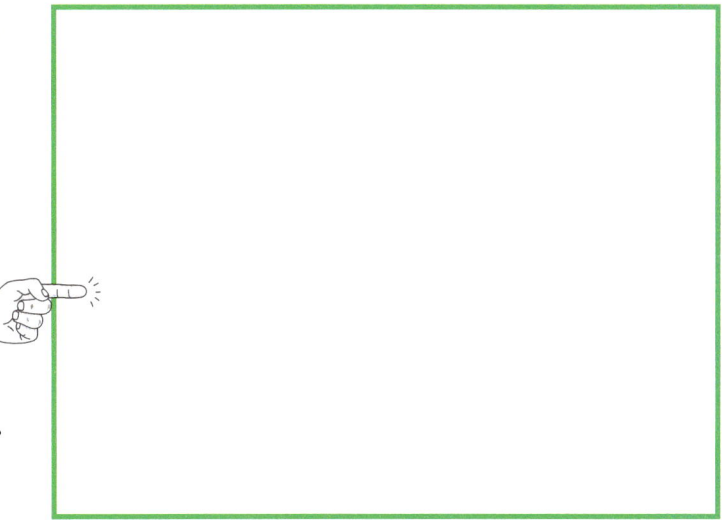

4. There are 12 darts players in a league.

   Each player plays every other player four times.

   Work out the total number of games played.

When we are given information about a rounded number it is possible to work out some information about its original size. For example:

- The **Lower Bound** (LB) is the smallest value that would round up to the number in question.
- The **Upper Bound** (UB) is the smallest value above the number that does NOT round to that number.

An error interval uses these bounds to state the range of values for the size of the original number:

Lower Bound ≤ original number < Upper Bound

The lower bound rounds to the original number so we use ≤

The upper bound does not round to the number so we use <

Sometimes we need to use our understanding of upper and lower bounds when solving problems ...

CHECK-IN

1. Calculate $813 \div 44$. Give your answer correct to two significant figures.
2. Calculate $(3.4 \times 10^5) \div (2.7 \times 10^3)$. Give your answer in standard form and correct to three significant figures.
3. A number N is rounded to 2.48 to three significant figures. What is the lowest possible value of N?
4. A number Y is rounded to 1300 to the nearest 100. Work out the difference between 1300 and the lowest possible value of Y.

---

e.g. 1a A rectangular field has a length of 180 metres and a width of 75 metres. Both measurements are given to the nearest five metres.

Work out an upper bound for the perimeter of the field.

❶ Consider the bounds of the rounded values

Error interval for length is $177.5 \text{ m} \leq \text{length} < 182.5 \text{ m}$

Error interval for width is $72.5 \text{ m} \leq \text{width} < 77.5 \text{ m}$

❷ Consider the required calculation

Perimeter = $2 \times \text{length} + 2 \times \text{width}$

❸ Choose the correct bounds: *The upper bound for perimeter needs both these values to be maximised*

Perimeter = $2 \times 182.5 + 2 \times 77.5$
= 520

❹ State the answer

Upper bound for perimeter = 520 metres

e.g. 1b Work out the lower bound for the area of the field.

Area = length × width

The lower bound for area needs both the length and width to be minimised

Area = 177.5×72.5
= 12 868.75   *Both lower bounds are needed*

Lower bound for area = $12\,868.75 \text{ m}^2$

e.g. 2 The distance from Earth to the moon is 384 400 km, to the nearest 100 kilometres.

To the nearest hour, astronauts on Apollo 11 took 76 hours to travel from Earth to the moon.

Calculate the lower bound for the average speed of their journey to the nearest whole number.

❶ Consider the bounds of the rounded values

Error interval for distance: $384\,350 \text{ km} \leq d < 384\,450 \text{ km}$

Error interval for time in hours: $75.5 \text{ hours} \leq t < 76.5 \text{ hours}$

❷ Average speed = $\dfrac{\text{Total distance}}{\text{Total time}}$ *The lower bound needs: this value minimised, and this value maximised*

❸ Average speed = $\dfrac{384\,350}{76.5}$ *These values give the smallest possible speed*

= 5024.183... = 5024 (nearest whole number)

❹ Lower bound for average speed = 5024 km/h

A. The lengths in the diagram have been rounded to the nearest centimetre.

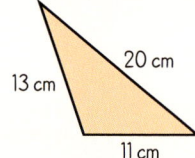

Calculate the lower bound of the perimeter of the triangle.

B. The lengths in the trapezium have been rounded to the nearest millimetre.

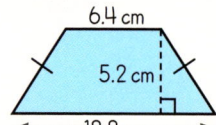

Calculate the upper bound for the area of the trapezium.

C. The population of Herefordshire is 192 000 to the nearest 1000.

The area of Herefordshire is 2200 km^2 to the nearest 100.

Calculate the upper bound for the population density of Herefordshire.

HINT: Population density is measured in people per square kilometre

GO! CALCULATING WITH BOUNDS

Champion!

1. Terry is told that:
   $a = 2.3$ to one decimal place
   $b = 4.28$ to two decimal places
   $C = b - a$

   He works out a lower bound of C as:
   $$4.275 - 2.25 = 2.025$$

   Explain why Terry is incorrect.

2. $m = 9.6$ to 2 significant figures
   $n = 13.76$ to 2 decimal places
   $p = 20$ to the nearest 10

   Work out the lower bound for B if:

   a) $B = m^2 - n^2$

   b) $B = \dfrac{m + n}{2p}$

   c) $B = \dfrac{m}{\sqrt{np}}$

   Give your answers to 2 significant figures.

3. An astronomical unit is 1.5×10^{11} metres to two significant figures.

   The speed of light is 3.00×10^8 metres per second to three significant figures.

   Calculate an upper bound for the length of time, in seconds, it takes light to travel one astronomical unit.

4. A block of concrete is in the shape of a cuboid. The block measures 15 cm by 45 cm by 20 cm, where all measurements are to the nearest millimetre.

   The mass of the block is 32.5 kg to three significant figures.

   By using bounds, work out the density of the block. Give your answer in g/cm³ and to a suitable degree of accuracy.

   Justify your answer.

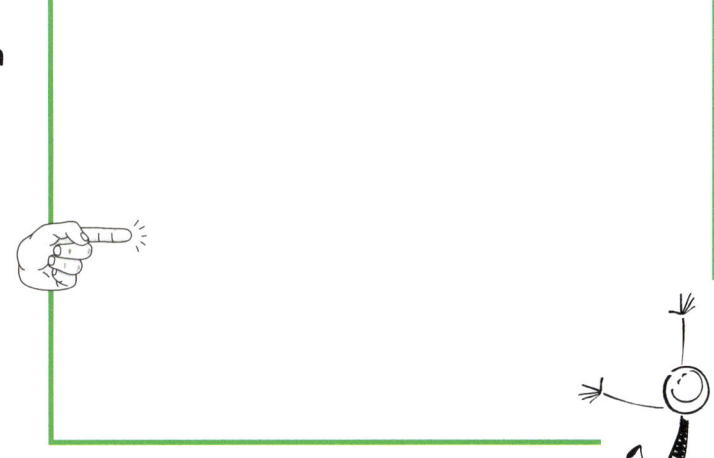

READY?

Sometimes we can work out roots exactly. For example:

$\sqrt{36} = 6 \qquad \sqrt[3]{8} = 2 \qquad \sqrt{\frac{1}{4}} = \frac{1}{2}$

But very often we can't find the exact answer. e.g.

$\sqrt{2} = 1.414\,213\,562\,373\,095\,048\,801\,688\,724\,209\,698\,078\ldots$

Roots that can't be worked out exactly are called **surds**.

We can simplify surds using:

Fact 1: $\sqrt{a \times b} = \sqrt{a} \times \sqrt{b}$

Fact 2: $\sqrt{\dfrac{a}{b}} = \dfrac{\sqrt{a}}{\sqrt{b}}$

√a + b is NOT the same as √a + √b

To simplify a surd involving a square root, we need to look for factors of the number that are also square numbers. For example, if we are simplifying $\sqrt{20}$

the factors of 20 are: 1, 2, **4**, 5, 10 and 20

↳ 4 is a square number

CHECK-IN

1. Write down the first 12 square numbers.

2. Find all the factors of 72 that are square numbers.

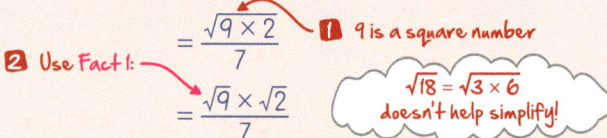

e.g. 1 Simplify $\sqrt{20}$

1. Look for the largest factor of 20 that is a square number
   4 is the largest square number factor

2. Use Fact 1 to break up 20 using the square number factor
   $\sqrt{20} = \sqrt{4 \times 5}$
   $= \sqrt{4} \times \sqrt{5}$

   It is also true that $\sqrt{20} = \sqrt{2 \times 10}$, but it doesn't help simplify!

3. Evaluate where possible: $= 2 \times \sqrt{5}$

4. Write as simply as possible: $= 2\sqrt{5}$

e.g. 2 Simplify fully $\sqrt{\dfrac{18}{49}}$

First use **Fact 2**: $\sqrt{\dfrac{18}{49}} = \dfrac{\sqrt{18}}{\sqrt{49}}$

Then deal with the numerator and denominator separately:

We can evaluate $\sqrt{49}$ → $\dfrac{\sqrt{18}}{\sqrt{49}}$ ← We need to simplify $\sqrt{18}$

2. Use **Fact 1**: $= \dfrac{\sqrt{9 \times 2}}{7}$ ← 9 is a square number

   $= \dfrac{\sqrt{9} \times \sqrt{2}}{7}$

   $\sqrt{18} = \sqrt{3 \times 6}$ doesn't help simplify!

3. Evaluate: $= \dfrac{3 \times \sqrt{2}}{7} = \dfrac{3\sqrt{2}}{7}$ 4. Simplify if possible

If a surd appears in the denominator of a fraction, we need to eliminate it. This is called **rationalising the denominator** of a fraction.

e.g. 3 Rationalise the denominator of $\dfrac{6}{\sqrt{7}}$

Multiply the numerator and denominator by the surd:

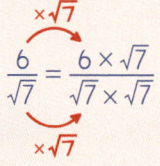

$\dfrac{6}{\sqrt{7}} = \dfrac{6 \times \sqrt{7}}{\sqrt{7} \times \sqrt{7}}$

Evaluate and write as simply as possible:

$\dfrac{6 \times \sqrt{7}}{\sqrt{7} \times \sqrt{7}} = \dfrac{6\sqrt{7}}{7}$

Remember:
$\sqrt{7} \times \sqrt{7} = 7^{\frac{1}{2}} \times 7^{\frac{1}{2}}$
$= 7^{\frac{1}{2}+\frac{1}{2}}$
$= 7^1$
$= 7$

SET?

A. Simplify fully:

(i) $\sqrt{50}$

(ii) $\sqrt{72}$

(iii) $\sqrt{32}$

(iv) $\sqrt{125}$

(v) $\sqrt{180}$

(vi) $\sqrt{108}$

B. Rationalise the denominator:

(i) $\dfrac{4}{\sqrt{3}}$

(ii) $\dfrac{8}{\sqrt{5}}$

C. Rationalise the denominator of these fractions, and simplify them fully:

(i) $\dfrac{9}{5\sqrt{7}}$

(ii) $\dfrac{4}{3\sqrt{2}}$

(iii) $\dfrac{24}{5\sqrt{6}}$

D. Simplify fully:

(i) $\sqrt{\dfrac{7}{25}}$

(ii) $\sqrt{\dfrac{13}{144}}$

(iii) $\sqrt{\dfrac{27}{64}}$

(iv) $\sqrt{\dfrac{20}{81}}$

Remember: choose the largest square number factor

GO! SURDS 1

Outstanding!

1. Vanessa is asked to simplify fully $\sqrt{48}$
   She writes:
   $$\sqrt{48} = \sqrt{4 \times 12} = \sqrt{4} \times \sqrt{12} = 2\sqrt{12}$$
   Do you agree with Vanessa? Explain why.

2. Show that $\sqrt{507}$ can be written in the form $k\sqrt{3}$ where k is an integer.
   State the value of k.

3. Find the exact value of x in this triangle:

   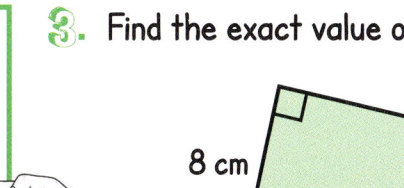

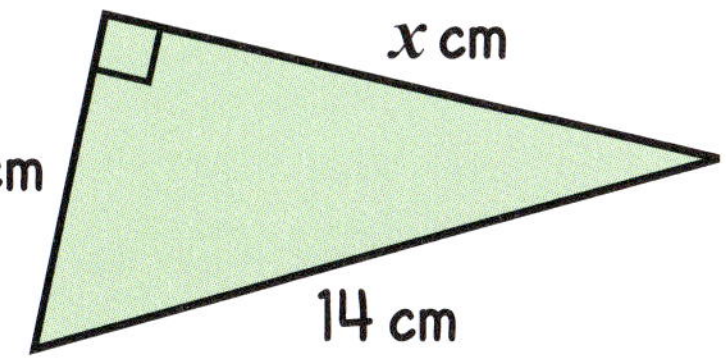

   Give your answer in its simplest form.

   HINT: Remember Pythagoras' theorem

4. Show that $\sqrt{\dfrac{28}{49}}$ can be written in the form $\dfrac{a\sqrt{b}}{7}$ where a and b are integers.

5. Place the digits 3, 4, 5 and 6 to make a correct statement.

   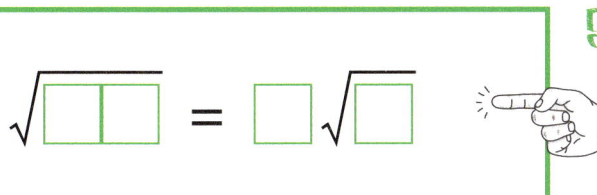

READY?

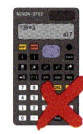

We have seen that we can simplify surds using:

$$\sqrt{a \times b} = \sqrt{a} \times \sqrt{b} \quad \text{and} \quad \sqrt{\frac{a}{b}} = \frac{\sqrt{a}}{\sqrt{b}}$$

We can also use these facts in more complex questions.

e.g. 1 Expand and simplify $\sqrt{5}(13 - \sqrt{35})$

A multiplication grid can be useful:

	13	$-\sqrt{35}$
$\sqrt{5}$	$13\sqrt{5}$	$-\sqrt{175}$

$\sqrt{5} \times 13$ $\quad$ $\sqrt{5} \times -\sqrt{35}$

Expand: $\sqrt{5}(13 - \sqrt{35}) = 13\sqrt{5} - \sqrt{175}$

Simplify:
$$= 13\sqrt{5} - \sqrt{25 \times 7}$$
$$= 13\sqrt{5} - \sqrt{25} \times \sqrt{7}$$
$$= 13\sqrt{5} - 5\sqrt{7}$$

TECHNICAL STUFF: $\sqrt{5}(13 - \sqrt{35})$ is equivalent to $\sqrt{5}(13 + -\sqrt{35})$

e.g. 2 Expand and simplify $(2 - \sqrt{6})(4 + \sqrt{15})$

	4	$\sqrt{15}$
2	8	$2\sqrt{15}$
$-\sqrt{6}$	$-4\sqrt{6}$	$-\sqrt{90}$

$2 \times \sqrt{15}$; $-\sqrt{6} \times \sqrt{15}$; $-\sqrt{6} \times 4$; 2×4

Expand:
$$(2 - \sqrt{6})(4 + \sqrt{15}) = 8 + 2\sqrt{15} - 4\sqrt{6} - \sqrt{90}$$

$\sqrt{90}$ can be simplified

$$\sqrt{90} = \sqrt{9 \times 10} = \sqrt{9} \times \sqrt{10} = 3 \times \sqrt{10} = 3\sqrt{10}$$

$$(2 - \sqrt{6})(4 + \sqrt{15}) = 8 + 2\sqrt{15} - 4\sqrt{6} - 3\sqrt{10}$$

CHECK-IN (JUSTAROO Airlines — Ready for take off?)

1 Simplify:

$\sqrt{8} =$ $\qquad$ $\sqrt{54} =$ $\qquad$ $\sqrt{288} =$

2 Rationalise the denominator and simplify:

$\frac{6}{\sqrt{5}} =$ $\qquad$ $\frac{7}{2\sqrt{3}} =$ $\qquad$ $\frac{8}{9\sqrt{2}} =$

The 'conjugate' of the expression $(x + \sqrt{y})$ is $(x - \sqrt{y})$. Something interesting happens when we multiply these expressions together ...

e.g. 3 Expand and simplify $(7 + \sqrt{3})(7 - \sqrt{3})$

	7	$-\sqrt{3}$
7	49	$-7\sqrt{3}$
$\sqrt{3}$	$7\sqrt{3}$	-3

These two terms cancel each other out

$$(7 + \sqrt{3})(7 - \sqrt{3}) = 49 - 7\sqrt{3} + 7\sqrt{3} - 3$$
$$= 46$$

Calculating with surds in this way can help us to rationalise the denominator in more complex cases ...

e.g. 4 Rationalise the denominator of $\frac{\sqrt{6} + \sqrt{2}}{8 - \sqrt{2}}$

1 Multiply the numerator and denominator by the conjugate of the denominator:

$$\frac{\sqrt{6} + \sqrt{2}}{8 - \sqrt{2}} = \frac{(\sqrt{6} + \sqrt{2})(8 + \sqrt{2})}{(8 - \sqrt{2})(8 + \sqrt{2})} \quad \times(8 + \sqrt{2})$$

2 Evaluate the numerator and denominator:

	8	$\sqrt{2}$
$\sqrt{6}$	$8\sqrt{6}$	$\sqrt{12}$
$\sqrt{2}$	$8\sqrt{2}$	2

$8\sqrt{6} + \sqrt{12} + 8\sqrt{2} + 2$ ($\sqrt{12} = 2\sqrt{3}$)

	8	$\sqrt{2}$
8	64	$8\sqrt{2}$
$-\sqrt{2}$	$-8\sqrt{2}$	-2

$64 + 8\sqrt{2} - 8\sqrt{2} - 2 = 62$

3 Simplify:

$$= \frac{8\sqrt{6} + 2\sqrt{3} + 8\sqrt{2} + 2}{62} = \frac{4\sqrt{6} + \sqrt{3} + 4\sqrt{2} + 1}{31}$$

SET?

A. Expand, and simplify if possible:

(i) $\sqrt{3}(4 + \sqrt{7})$

(ii) $\sqrt{2}(9 - \sqrt{30})$

(iii) $\sqrt{3}(\sqrt{2} + 24)$

(iv) $\sqrt{10}(\sqrt{7} - \sqrt{5})$

(v) $\sqrt{3}(\sqrt{15} + \sqrt{39})$

B. Expand, and simplify:

(i) $(3 + \sqrt{6})(4 + \sqrt{6})$

(ii) $(7 + \sqrt{3})(5 - \sqrt{6})$

(iii) $(10 - \sqrt{14})(\sqrt{2} - 1)$

(iv) $(12 + 2\sqrt{3})(8 + \sqrt{3})$

(v) $(5 + \sqrt{2})^2$

C. Rationalise the denominator of these fractions, and simplify them fully:

(i) $\frac{3 + \sqrt{6}}{4 - \sqrt{6}}$

(ii) $\frac{7 + \sqrt{3}}{5 + \sqrt{6}}$

(iii) $\frac{12 + 2\sqrt{3}}{8 - \sqrt{3}}$

GO! SURDS 2

Excellent!

1. Dylan writes:
   $$(3a - \sqrt{b})^2 = 9a^2 - b^2$$
   Show that Dylan is not correct.

2. Find values for p, q and r to make this statement true:
   $$(p + \sqrt{7})(8 - \sqrt{q}) = 17 + r\sqrt{7}$$

   HINT: Start with q

3. Find the values of b and c given that:
   $$x^2 + bx + c = (x + (7 + \sqrt{5}))(x + (7 - \sqrt{5}))$$

4. The Golden Ratio is the value Φ where:
   $$\Phi = \frac{1 + \sqrt{5}}{2}$$
   Show that:
   a) $\Phi^2 = \Phi + 1$
   b) $\frac{1}{\Phi} = \Phi - 1$

READY?

A mathematical statement involving a combination of numbers, symbols and/or operations is called an **expression**. The different parts of an expression have names ...

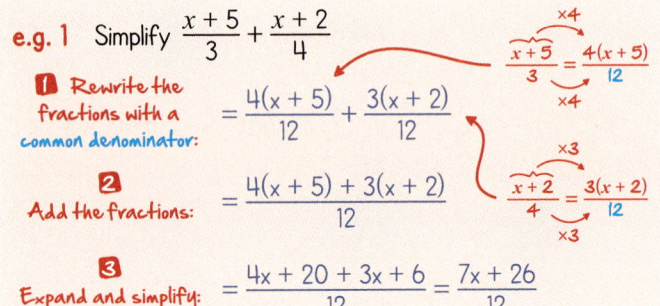

An expression does NOT include an equals symbol

CHECK-IN

1 Calculate:

$\frac{2}{7} + \frac{1}{8} =$ $\frac{7}{9} - \frac{5}{12} =$

2 Expand and simplify:

$3(2x + 4) + 2(x + 3)$

$9(x - 5) - 2(3x + 7)$

Some expressions include algebraic fractions. We can simplify these expressions using the same approaches as when calculating with fractions.

e.g. 1 Simplify $\frac{x+5}{3} + \frac{x+2}{4}$

1 Rewrite the fractions with a common denominator:
$= \frac{4(x+5)}{12} + \frac{3(x+2)}{12}$

2 Add the fractions:
$= \frac{4(x+5) + 3(x+2)}{12}$

3 Expand and simplify:
$= \frac{4x + 20 + 3x + 6}{12} = \frac{7x + 26}{12}$

$\frac{x+5}{3} \xrightarrow{\times 4} \frac{4(x+5)}{12}$

$\frac{x+2}{4} \xrightarrow{\times 3} \frac{3(x+2)}{12}$

Sometimes there may be variables in the denominator:

e.g. 2 Write $\frac{4}{n+5} + \frac{2}{n-3}$ as a single fraction.

1 Rewrite with a common denominator:
$\frac{4(n-3)}{(n+5)(n-3)} + \frac{2(n+5)}{(n-3)(n+5)}$

2 Add ...
$= \frac{4(n-3) + 2(n+5)}{(n+5)(n-3)}$

3 Expand ...
$= \frac{4n - 12 + 2n + 10}{(n+5)(n-3)} = \frac{6n - 2}{(n+5)(n-3)}$

$\frac{4}{n+5} \xrightarrow{\times(n-3)} \frac{4(n-3)}{(n+5)(n-3)}$

$\frac{2}{n-3} \xrightarrow{\times(n+5)} \frac{2(n+5)}{(n-3)(n+5)} = \frac{2(n+5)}{(n+5)(n-3)}$

No need to factorise or expand brackets

e.g. 3 Write $\frac{3y}{y+2} - \frac{y}{y-5}$ as a single fraction.

1 Rewrite:
$\frac{3y(y-5)}{(y+2)(y-5)} - \frac{y(y+2)}{(y-5)(y+2)}$

2 Subtract the fractions:
$= \frac{3y(y-5) - y(y+2)}{(y+2)(y-5)}$

3 Expand and simplify:
$= \frac{3y^2 - 15y - y^2 - 2y}{(y+2)(y-5)}$

$= \frac{2y^2 - 17y}{(y+2)(y-5)}$

No need to factorise or expand brackets

Further manipulation is sometimes needed ...

e.g. 4 Show that $\frac{4x+3}{5} - \frac{x}{8}$ can be written in the form $\frac{a(bx+c)}{d}$ where a, b, c and d are integers.

1 Rewrite:
$\frac{4x+3}{5} - \frac{x}{8} = \frac{8(4x+3)}{40} - \frac{5x}{40}$

2 Subtract:
$= \frac{8(4x+3) - 5x}{40}$

3 Expand and simplify:
$= \frac{32x + 24 - 5x}{40} = \frac{27x + 24}{40}$

4 Factorise the numerator to write in the required form:
$= \frac{3(9x+8)}{40}$

A. Simplify:

(i) $\frac{x+1}{6} + \frac{x+3}{5}$

(ii) $\frac{n-3}{3} + \frac{3n+2}{7}$

(iii) $\frac{2a-1}{2} - \frac{a+3}{5}$

B. Write $\frac{x}{4} - \frac{x-3}{6}$ as a single fraction in its simplest form.

C. Write as a single fraction:

(i) $\frac{1}{x+4} + \frac{3}{x+2}$

(ii) $\frac{8}{y+5} - \frac{3}{y}$

(iii) $\frac{5}{2p-3} - \frac{2}{p+6}$

(iv) $\frac{5x}{x-4} + \frac{2x}{x-3}$

SET?

D. Show that $\frac{x+3}{4} + \frac{3x-1}{3}$ can be written in the form $\frac{a(bx+c)}{d}$ where a, b, c and d are integers.

26

GO!

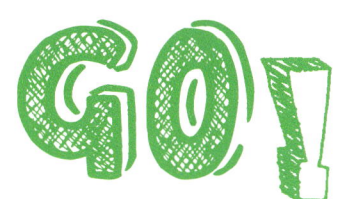

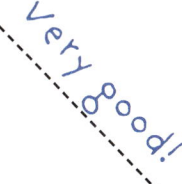

ALGEBRAIC FRACTIONS 1

Very good!

1. Write $\frac{4}{m-n} - \frac{5}{n}$ as a single fraction in its simplest form.

2. Place the digits 2, 3, 4 and 5 to make the statement correct.

$$\frac{\square x}{x+4} + \frac{x}{x+\square} = \frac{x(\square x + 14)}{(x+\square)(x+5)}$$

3. Mark is asked to write $\frac{3a+2}{2} - \frac{4a-1}{3}$ as a single fraction. He writes:

$$\frac{3a+2}{2} - \frac{4a-1}{3}$$

$$= \frac{3(3a+2) - 2(4a-1)}{6}$$

$$= \frac{9a+6-8a-2}{6}$$

$$= \frac{a+4}{6}$$

Explain why Mark's solution is not correct.

4. Write $\frac{2(x+4)}{5} + \frac{2x}{3} - \frac{3(x+1)}{2}$ as a fraction in the form $\frac{ax+b}{c}$ where a, b and c are integers.

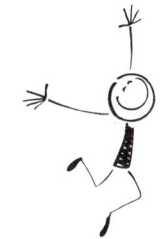

A mathematical statement containing an equals symbol is called an **equation**. The different parts of an equation have names...

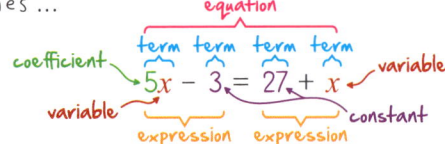

This equation includes one variable (x). The variable is called an **unknown**. We can often solve equations to find the value of an unknown. The equation must stay balanced. This means applying the same operation (such as $+$, $-$, $\times$, $\div$, or $\sqrt{\ }$) to each side. For example:

Subtract x from both sides: $5x - 3 = 27 + x$, $-x$, $-x$
Add 3 to both sides: $4x - 3 = 27$, $+3$, $+3$
Divide both sides by 4: $4x = 30$, $\div 4$, $\div 4$
$x = \frac{30}{4} = \frac{15}{2}$ or $7\frac{1}{2}$ or 7.5

We may need to give the answer as a fraction, mixed number, or a decimal

e.g. 1 Solve $x = \frac{3x + 2}{8}$

1 It is often helpful to eliminate fractions from the equation:
$x = \frac{3x+2}{8}$, $\times 8$, $\times 8$ — Multiply both sides by 8

2 Make an equation with the unknown on one side only:
$8x = 3x + 2$, $-3x$, $-3x$ — Subtract $3x$ from both sides
$5x = 2$
$\div 5$, $\div 5$ — Divide both sides by 5

3 Isolate the unknown $\rightarrow$ $x = \frac{2}{5}$

So $x = \frac{2}{5}$ is the solution

CHECK-IN

Solve:
1. $3x + 17 = 2$
2. $4y + 1 = 13 - 2y$
3. $\frac{5m + 1}{2} = 8$
4. $\frac{3(k - 4)}{2} = 6$

Ready for take off?

e.g. 2 Solve $5 + \frac{2}{3n} = 8$

1 Eliminate any fractions (multiply every term by $3n$)
$5 + \frac{2}{3n} = 8$, $\times 3n$, $\times 3n$
$15n + 2 = 24n$

2 Make the unknown appear on one side only:
$-15n$, $-15n$
$2 = 9n$
$\div 9$, $\div 9$

3 Isolate the unknown $\rightarrow$ $\frac{2}{9} = n$

so $n = \frac{2}{9}$ is the solution

OR We could start by subtracting 5 from both sides...
$5 + \frac{2}{3n} = 8$, -5, -5
$\frac{2}{3n} = 3$
$\times 3n$, $\times 3n$
$2 = 9n$
$\div 9$, $\div 9$

Sometimes we need to manipulate one or both sides of the equation...

e.g. 3 Find the value of y if $4(2y - 7) = \frac{7 - 5y}{2}$

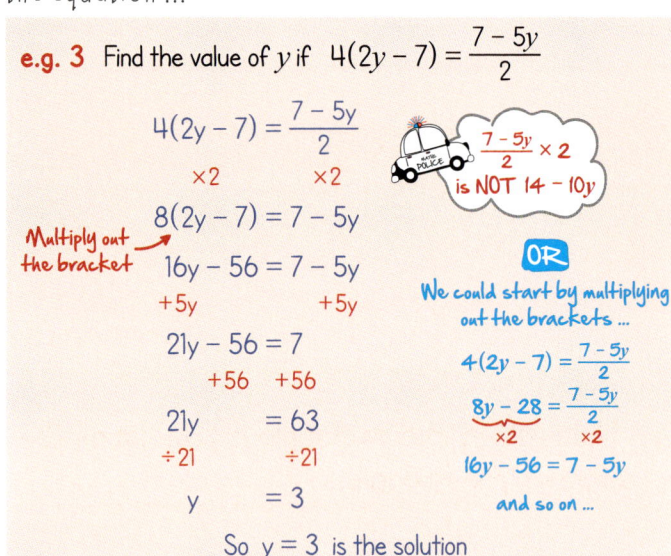

$4(2y - 7) = \frac{7 - 5y}{2}$
$\times 2$, $\times 2$
Multiply out the bracket $\rightarrow$ $8(2y - 7) = 7 - 5y$
$16y - 56 = 7 - 5y$
$+5y$, $+5y$
$21y - 56 = 7$
$+56$, $+56$
$21y = 63$
$\div 21$, $\div 21$
$y = 3$

$\frac{7 - 5y}{2} \times 2$ is NOT $14 - 10y$

OR We could start by multiplying out the brackets...
$4(2y - 7) = \frac{7 - 5y}{2}$
$8y - 28 = \frac{7 - 5y}{2}$
$\times 2$, $\times 2$
$16y - 56 = 7 - 5y$
and so on...

So $y = 3$ is the solution

SET?

A. Solve $x = \frac{7x - 2}{3}$

B. Solve $3y = \frac{5 - 2y}{6}$

C. Solve $\frac{12m + 7}{4} = m$

D. Solve $2(3x - 5) = \frac{10x + 2}{3}$

E. Solve $6w + 4 = \frac{15w + 6}{2}$

F. Solve $\frac{8a + 9}{2} = 3 + 2a$

G. Solve $\frac{3}{4p} + 2 = \frac{9}{4}$

H. Solve $-6 = 1 - \frac{7}{2n}$

I. Solve $7 + \frac{4}{5v} = 2$

Give all solutions in their simplest form

GO! SOLVING HARDER EQUATIONS

Brilliant!

1. $\dfrac{3}{5n} + \dfrac{5}{2n} = 93$

   Find the value of n

2. Solve:

   a) $\dfrac{4a - 9}{5} + \dfrac{7(15 - 2a)}{3} = 10$

   b) $3b + \dfrac{8(b - 3)}{9} = \dfrac{2}{3}$

3. The area of the rectangle is four times the area of the triangle.

   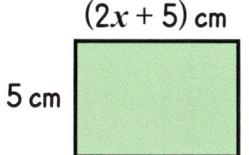
   (2x + 5) cm, 5 cm

   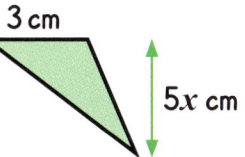
   3 cm, 5x cm

   a) Work out the value of x
   b) Find the value of the longest side of the rectangle.

4. The graphs of $y = 3(2x + 1)$ and $y = \dfrac{2x + 11}{2}$ intersect at the point A.

   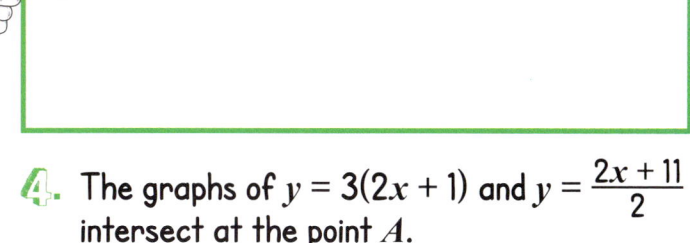

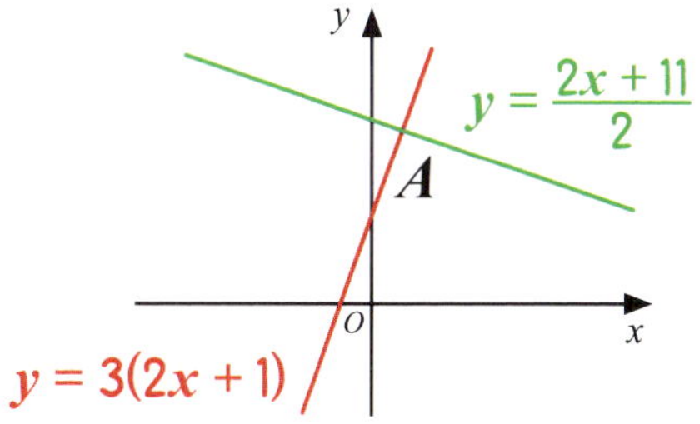

   Work out the coordinates of the point A.

A **quadratic expression** is one of the form $ax^2 + bx + c$. a, b and c can be any number. For example:

$x^2 + 5x + 6$ ($a = 1, b = 5$ and $c = 6$)

$2x^2 - 4x + 1$ ($a = 2, b = -4$ and $c = 1$)

$8x^2 + 5$ ($a = 8, b = 0$ and $c = 5$)

TECHNICAL STUFF: $2x^2 - 4x + 1 = 2x^2 + -4x + 1$

Some quadratic expressions can be factorised. This is the opposite of expanding:

$$x^2 + 5x + 6 = (x + 2)(x + 3)$$

FACTORISING → / ← EXPANDING

The multiplication grid for expanding brackets is also helpful when factorising ...

e.g. 1 Factorise $2x^2 + 11x + 15$

1. Place the quadratic and constant terms in the grid:
2. Fill in any information which is certain:
3. Consider the factor pairs of the constant term.
   1 and 15, 3 and 5, −1 and −15 or −3 and −5
   Experiment with different pairs:

   $6x + 5x = 11x$ ✓

4. When the pair of correct terms is found, write the solution:
   $2x^2 + 11x + 15 = (x + 3)(2x + 5)$

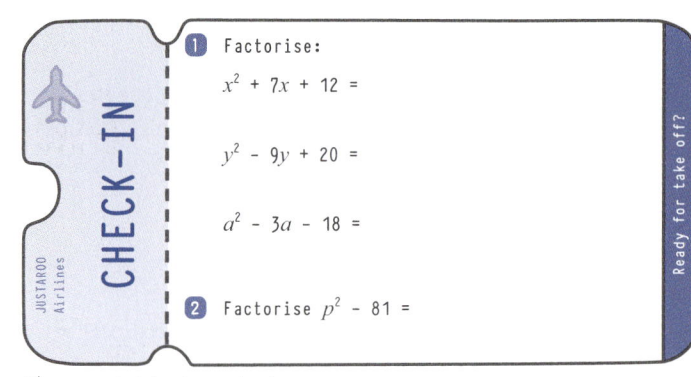

CHECK-IN

1. Factorise:
   $x^2 + 7x + 12 =$
   $y^2 - 9y + 20 =$
   $a^2 - 3a - 18 =$

2. Factorise $p^2 - 81 =$

Ready for take off?

The variable is not always x ...

e.g. 2 Factorise $3p^2 + 17p - 6$

1. [grid with $3p^2$ and -6]
2. [grid with $3p$, p, $3p^2$, -6]

3. The factor pairs of −6 are:
   −2 and 3, 2 and −3, −1 and 6, 1 and −6

   $18p + -1p = 17p$ ✓

4. $3p^2 + 17p - 6 = (p + 6)(3p - 1)$

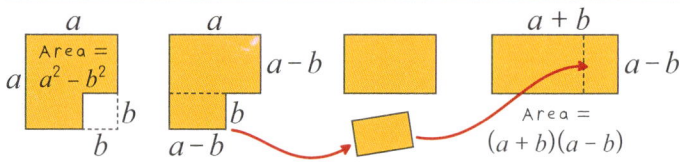

Look out for 'the difference of two squares' ...

e.g. 3 Factorise $16x^2 - 49$

The difference of two squares tells us that: $a^2 - b^2 = (a + b)(a - b)$

$\sqrt{16x^2} = 4x$ and $\sqrt{49} = 7$

so $16x^2 - 49 = (4x + 7)(4x - 7)$

A. Factorise:

(i) $2x^2 + 11x + 14$

(ii) $2y^2 + 15y + 27$

(iii) $3x^2 + 7x + 4$

(iv) $11p^2 + 46p + 8$

B. Factorise:

(i) $3x^2 + 13x - 10$

(ii) $2n^2 - 5n - 12$

(iii) $5w^2 - 13w + 6$

(iv) $7x^2 - 34x - 5$

C. Factorise:

(i) $25x^2 - 36$

(ii) $144c^2 - 121$

30

GO! FACTORISING

Blinding!

1. Steve is given the quadratic expression:
   $$3y^2 + 19y + 28$$
   He says:
   "You cannot factorise the expression as there aren't two numbers with a product of 28 that add to make 19"

   Explain the mistake that Steve has made.

2. Factorise $9a^4 - 4b^2$

3. Factorise fully $6x^2 - 26x + 24$

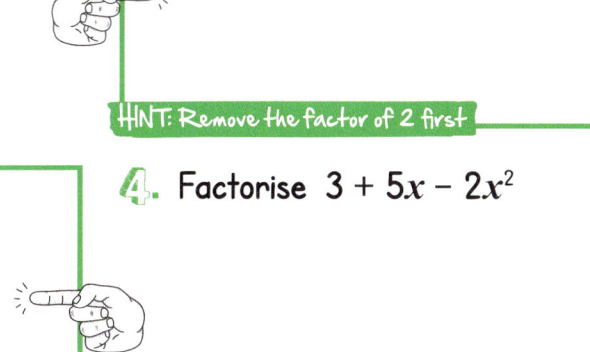

   HINT: Remove the factor of 2 first

4. Factorise $3 + 5x - 2x^2$

5. Place the digits 1, 2, 3, 4 and 5 to complete the statement.

   $$2x^2 - \square x - \square 2$$
   $$=$$
   $$(\square x + \square)(x - \square)$$

An algebraic expression with two terms is sometimes called a binomial. For example:

$x + 5 \qquad 3x - 1 \qquad -7 - 3y \qquad 8a + 5b$

We can multiply two binomials together. It can help to use a multiplication grid to keep track of all the steps:

$(x + 5)(3x - 1)$
$= 3x^2 + 15x - x - 5$
$= 3x^2 + 14x - 5$

×	$3x$	-1
x	$3x^2$	$-x$
5	$15x$	-5

TECHNICAL STUFF: $3x - 1$ is equivalent to $3x + -1$

The multiplication grid is very helpful when multiplying three binomials (sometimes called triple brackets)...

e.g. 1 Expand and simplify $(x + 3)(x + 2)(x + 5)$

1 Expand and simplify the first two binomials

$(x + 3)(x + 2) = x^2 + 3x + 2x + 6$
$= x^2 + 5x + 6$

×	x	2
x	x^2	$2x$
3	$3x$	6

2 Multiply this quadratic expression by the third binomial

$(x + 3)(x + 2)(x + 5) = (x^2 + 5x + 6)(x + 5)$

×	x^2	$5x$	6
x	x^3	$5x^2$	$6x$
5	$5x^2$	$25x$	30

3 Simplify
$x^3 + 5x^2 + 5x^2 + 25x + 6x + 30$
$= x^3 + 10x^2 + 31x + 30$

A cubic expression is one of the form $ax^3 + bx^2 + cx + d$ where a, b, c and d are integers

CHECK-IN — Expand and simplify:

1. $(x + 7)(x + 4) =$
2. $(x - 8)(x + 3) =$
3. $(y - 5)(y - 2) =$
4. $(2x + 1)(x - 6) =$
5. $(2p + 7)(3p + 9) =$

e.g. 2 Expand and simplify $(p - 4)(3p + 1)(p - 7)$

1
$(p - 4)(3p + 1) = 3p^2 - 12p + p - 4$
$= 3p^2 - 11p - 4$

×	$3p$	1
p	$3p^2$	p
-4	$-12p$	-4

2 $(p - 4)(3p + 1)(p - 7) = (3p^2 - 11p - 4)(p - 7)$

×	$3p^2$	$-11p$	-4
p	$3p^3$	$-11p^2$	$-4p$
-7	$-21p^2$	$77p$	28

3 $3p^3 - 21p^2 - 11p^2 + 77p - 4p + 28$
$= 3p^3 + 32p^2 + 73p + 28$

e.g. 3 Show that $(4x + 3)(3x - 8)(x + 6)$ can be written in the form $ax^3 + bx^2 + cx + d$ where a, b, c and d are integers.

1

×	$3x$	-8
$4x$	$12x^2$	$-32x$
3	$9x$	-24

$9x + -32x = -23x$

2

×	$12x^2$	$-23x$	-24
x	$12x^3$	$-23x^2$	$-24x$
6	$72x^2$	$-138x$	-144

3 $12x^3 + 72x^2 - 23x^2 - 138x - 24x - 144$
$= 12x^3 + 49x^2 - 162x - 144$

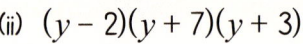

A. Expand and simplify:

(i) $(x + 4)(x + 1)(x + 5)$

(ii) $(y - 2)(y + 7)(y + 3)$

(iii) $(x + 5)(x - 2)(x - 8)$

B. Expand and simplify:

(i) $(2x + 1)(x + 3)(x + 8)$

(ii) $(p + 7)(5p - 2)(p + 4)$

(iii) $(3n - 2)(n - 4)(7n - 9)$

C. Show that $(5x + 2)(2x - 7)(3x + 8)$ can be written in the form $ax^3 + bx^2 + cx + d$ where a, b, c and d are integers:

32

Groovy!

GO!

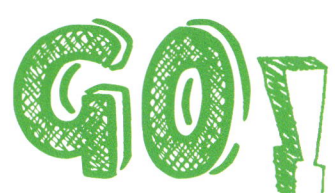

EXPANDING BRACKETS

1. Place the digits 2, 3, 4, 5 and 6 to complete the statement.

$$2x^3 - 15x^2 + 1\square x + \square 0$$
$$=$$
$$(\square x + 3)(x - \square)(x - \square)$$

2. Expand and simplify $(n + 2)(3n - 4)^2$

3. Expand and simplify:

$$(3k - 2)(3k + 2)(9k^2 + 4)$$

4. Expand and simplify $(2x + y)^3$

5. The volume of the cuboid is 1716 cm^3

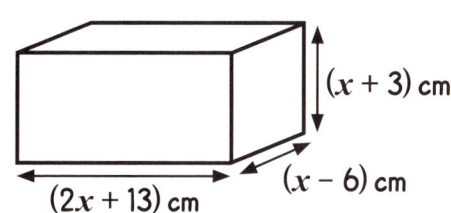

Show that $2x^3 + 7x^2 - 75x = 1950$

READY?

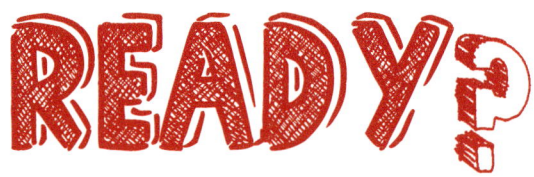

We have already seen how to add and subtract algebraic fractions. Sometimes we need to simplify algebraic fractions involving quadratic expressions.

e.g. 1 Simplify $\frac{6x + 24}{x^2 - 3x - 28}$

1 Factorise the numerator and denominator:

The HCF of 6 and 24 is 6, so use 6(.........)

$= \frac{6(x+4)}{(x+4)(x-7)}$

×	x	−7
x	x^2	−7x
4	4x	−28

2 Simplify by cancelling any common factors: $= \frac{6\cancel{(x+4)}}{\cancel{(x+4)}(x-7)}$

3 Write the simplified fraction: $= \frac{6}{x-7}$ No brackets needed!

As simplifying usually requires factorising, it is helpful to look for clues such as 'the difference of two squares'.

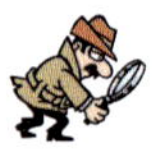

$\frac{2x^2 + 13x - 24}{4x^2 - 9}$ $4x^2 - 9 = (2x+3)(2x-3)$

e.g. 2 Simplify fully $\frac{2x^2 + 13x - 24}{4x^2 - 9}$

1a Factorise the denominator using the difference of two squares:

$= \frac{2x^2 + 13x - 24}{(2x+3)(2x-3)}$

1b Factorise the numerator: $= \frac{(x+8)(2x-3)}{(2x+3)(2x-3)}$

×	2x	−3
x	$2x^2$	−3x
8	16x	−24

2 Simplify by cancelling any common factors: $= \frac{(x+8)\cancel{(2x-3)}}{(2x+3)\cancel{(2x-3)}}$

3 Write the simplified fraction: $= \frac{x+8}{2x+3}$

CHECK-IN JUSTAROO Airlines

1 Write as a single fraction:

$\frac{x-6}{2} + \frac{x+3}{5}$

$\frac{2x}{x+3} + \frac{x}{x-1}$

2 Factorise:

$3x^2 + 7x + 4$

$4x^2 - 81$

Ready for take off?

Further manipulation is sometimes needed ...

e.g. 3 Show that $\frac{8y - 24}{y^2 - 8y + 15} \div \frac{y+5}{y^3 - 25y} = ay$ where a is an integer.

Rewrite as a multiplication: $= \frac{8y-24}{y^2-8y+15} \times \frac{y^3 - 25y}{y+5}$

$y^3 - 25y = y(y^2 - 25) = y(y+5)(y-5)$

Factorised: $= \frac{8(y-3)}{(y-5)(y-3)} \times \frac{y(y+5)(y-5)}{y+5}$

$= \frac{8\cancel{(y-3)}y\cancel{(y+5)}\cancel{(y-5)}}{\cancel{(y-5)}\cancel{(y-3)}\cancel{(y+5)}}$ Cancel all common factors

$= 8y$

e.g. 4 Show that $1 + \frac{8n + 32}{n^2 + 3n - 4}$ can be written in the form $\frac{n+a}{n+b}$ where a and b are integers.

$= 1 + \frac{8(n+4)}{(n+4)(n-1)}$ Factorised

$= 1 + \frac{8\cancel{(n+4)}}{\cancel{(n+4)}(n-1)} = 1 + \frac{8}{n-1}$ ← Simplified

$1 = \frac{n-1}{n-1}$: $= \frac{n-1}{n-1} + \frac{8}{n-1}$

$= \frac{n-1+8}{n-1} = \frac{n+7}{n-1}$

SET?

A. Simplify fully:

(i) $\frac{3x + 15}{x^2 + 3x - 10}$

(ii) $\frac{4n^2 + 12n}{n^2 - 3n - 18}$

(iii) $\frac{9x^2 - 16}{3x^2 + 22x + 24}$

(iv) $\frac{x^2 + 8x + 12}{3x^2 + x - 10}$

B. Show that each expression can be written in the form $\frac{ax + b}{x + c}$ where a, b and c are integers:

(i) $\frac{9x^2 - 36}{x^2 + 5x - 14} - 2$

(ii) $\frac{4x}{x-2} - \frac{3+x}{x^2 + x - 6}$

C. Write $\frac{3k^2 - 48}{k+4} \div \frac{k^2 - 9k + 20}{6k}$ as a single fraction in its simplest form.

Clever!

GO! ALGEBRAIC FRACTIONS 2

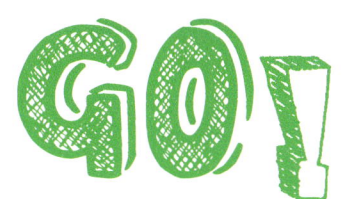

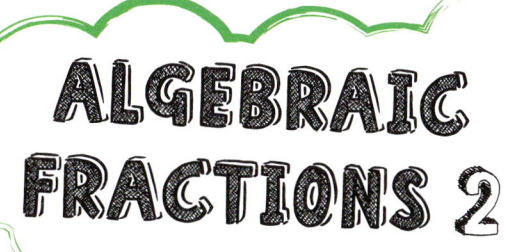

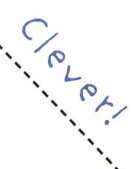

1. Steph is asked to simplify $\dfrac{y^2 + 13y + 36}{2y^2 + 11y + 12}$ fully. She writes:

$$= \frac{(y+4)(y+9)}{(y+4)(2y+3)}$$

$$= \frac{y+9}{2y+3}$$

$$= \frac{9}{2+3} = \frac{9}{5}$$

Explain why Steph's solution is not correct.

2. Simplify fully $\dfrac{5 - x}{x^2 + x - 30}$

3. Write

$$5 + \left[(x+4) \div \frac{2x^2 + 9x + 4}{x + 2}\right]$$

in the form $\dfrac{ax + b}{cx + d}$ where a, b, c and d are integers.

4. Write

$$\frac{4(4w+5)}{3w - 18} - \left[(w+3) \div \frac{3w^2 - 9w - 54}{4w - 1}\right]$$

as a single fraction in its simplest form.

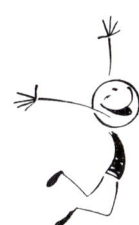

By adding consecutive integers, starting at 1, we can find the triangle numbers ...

  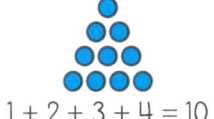

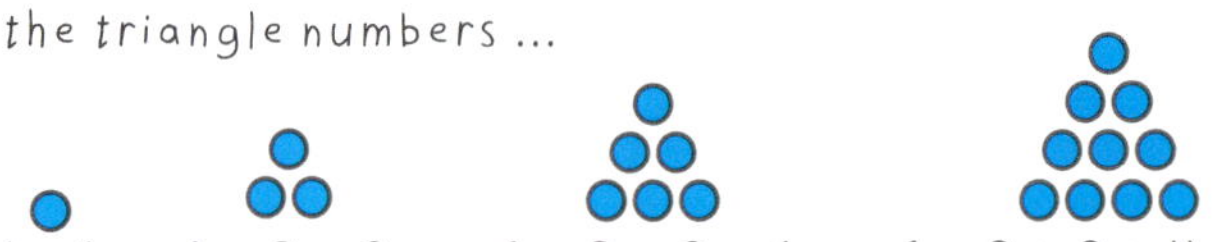

$1 = 1$ $1 + 2 = 3$ $1 + 2 + 3 = 6$ $1 + 2 + 3 + 4 = 10$

The nth triangle number is given by the expression:
$$\tfrac{1}{2}n(n+1)$$

For example ...

$\tfrac{1}{2} \times 1 \times 2 = 1$ $\tfrac{1}{2} \times 2 \times 3 = 3$ $\tfrac{1}{2} \times 3 \times 4 = 6$

 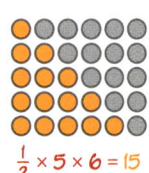

$\tfrac{1}{2} \times 4 \times 5 = 10$ $\tfrac{1}{2} \times 5 \times 6 = 15$

$\tfrac{1}{2}n(n+1)$ is half the product of any two consecutive numbers ... n and $(n+1)$. We can find expressions for other sets of numbers:

e.g. 1 Write expressions for two **consecutive** even numbers.

Even numbers are multiples of 2 → $2n$ and $2n + 2$ *The next even number is 2 greater*

e.g. 2 Write expressions for **any** two odd numbers.

An odd number must be one greater than an even number:

$2n + 1$ and $2p + 1$

Any letter can be used for the variable

$2n+1$ and $2n+3$ are NOT any two odd numbers (they are consecutive)

CHECK-IN

Expand and simplify:

1. $6(3n - 1) + 2(n - 5)$
2. $8(p + 4) - 3(4p - 2)$
3. $(2m + 1)(m + 3)$
4. $(2p + 1)(2p + 3)$
5. $(2a + 1)^2$

We can use these expressions in algebraic proofs ...

e.g. 3 Prove algebraically that the sum of the squares of any two even integers is a multiple of 4.

$2m$ and $2n$ are two even integers   *'Sum' means total*

$(2m)^2 + (2n)^2$
$= 4m^2 + 4n^2$
$= 4(m^2 + n^2)$ ← *Remove a factor of 4*

Always write a conclusion

So the sum of the squares of two even integers is a multiple of 4

Expanding and simplifying is often needed ...

e.g. 4 Prove that $5(3p - 5) + 7(p + 7)$ is even for all integer values of p.

Expand the brackets: $15p - 25 + 7p + 49$
Simplify: $= 22p + 24$
 $= 2(11p + 12)$ ← *Even numbers are multiples of 2*
Conclusion:
So $5(3p - 5) + 7(p + 7)$ is even for all integer values of p

Sometimes we can analyse the way numbers 'behave':

e.g. 5 Show that $n^2 + n$ is even for all integer values of n.

$n^2 + n = n(n + 1)$

If n is even: $n(n + 1) = $ even $\times$ odd $=$ even

If n is odd: $n(n + 1) = $ odd $\times$ even $=$ even

Conclusion: So $n^2 + n$ is even for all integer values of n

A. Write expressions for:

(i) Two consecutive odd numbers

(ii) A square number

(iii) Two consecutive square numbers

(iv) An odd square number

(v) Any two triangle numbers

B. Prove algebraically that the sum of four consecutive integers is always even.

C. Prove algebraically that

$$(2p)^2 + 4 + (2p + 4)^2$$

is a multiple of 4 for all integer values of p.

D. Prove algebraically that

$$5(3x - 2) + 2(9x - 6)$$

is a multiple of 11 for all integer values of x.

E. Prove that

$$(4n + 3)^2 + (3n + 4)^2 - n^2$$

is odd for all integer values of n.

F. Explain why the product of three consecutive integers is always even.

36

GO! ALGEBRAIC PROOF

1. Prove algebraically that the difference between
$$(6y + 5)^2 \text{ and } (6y - 5)^2$$
is a multiple of 24 for all integer values of y.

2. Prove algebraically that the cube of an odd number is odd.

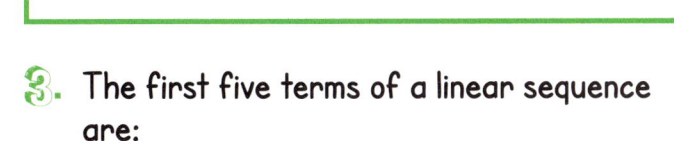

3. The first five terms of a linear sequence are:
$$3, \ 7, \ 11, \ 15, \ 19$$
Prove that the difference between the squares of any two consecutive terms of the sequence is always a multiple of 8

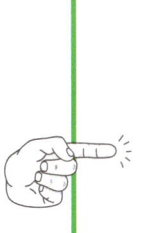

4. Prove algebraically that the sum of
$$\tfrac{1}{2}(k + 1)(k + 2) \text{ and } \tfrac{1}{2}(k + 2)(k + 3)$$
is always a square number.

Hats off!

READY?

Mathematical operations (such as ×3 and +5) can be used to build a **function**. The function then gives an **output** for any **input** we enter. A **function machine** can be used to help visualise this information. For example:

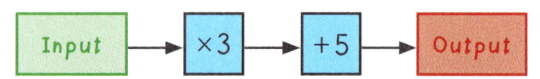

Input → ×3 → +5 → Output

A function needs one output for every possible input:

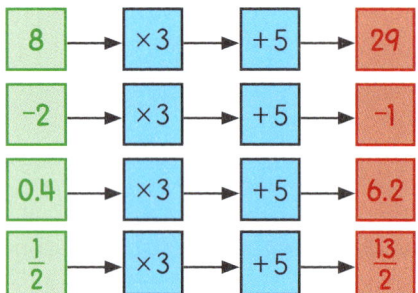

8 → ×3 → +5 → 29
−2 → ×3 → +5 → −1
0.4 → ×3 → +5 → 6.2
$\frac{1}{2}$ → ×3 → +5 → $\frac{13}{2}$

In general:

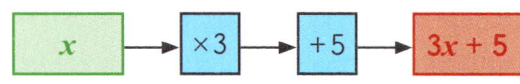

x → ×3 → +5 → $3x+5$

Using function notation, we write $f(x) = 3x+5$

e.g. 1 Given that $f(x) = 4x+7$ work out:

a) $f(5)$
$f(5) = 4 \times 5 + 7$
$= 20 + 7$
$= 27$

f(5) means that the input is 5

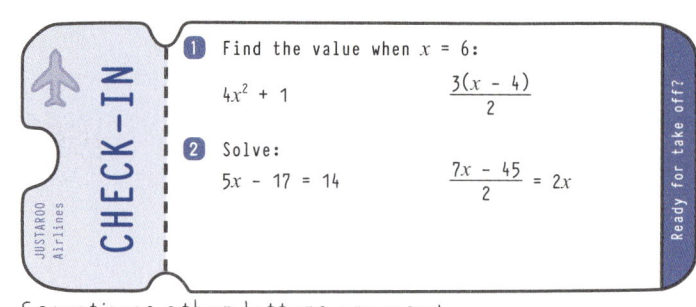

b) $f(-3)$
$= 4 \times -3 + 7$
$= -12 + 7$
$= -5$

f(x) can be read as "f of x"

CHECK-IN

1. Find the value when $x = 6$:
   $4x^2 + 1$
   $\frac{3(x-4)}{2}$

2. Solve:
   $5x - 17 = 14$
   $\frac{7x - 45}{2} = 2x$

Sometimes other letters are used.

e.g. 2 Given that $g(x) = 3x^2 - 4$ find $g(0.4)$

$g(0.4) = 3 \times 0.4^2 - 4$
$= 3 \times 0.16 - 4$
$= -3.52$

We can solve other problems using functions…

e.g. 3 $f(x) = 12 - 5x$
Find the value of x given that $f(x) = -43$

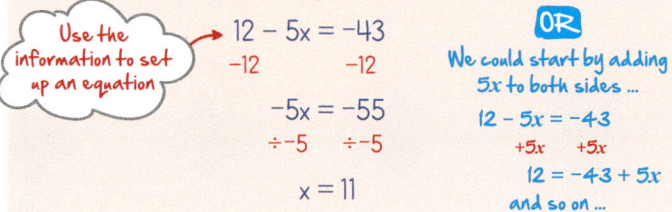

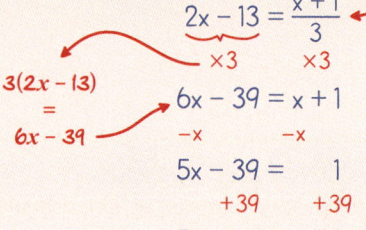

Use the information to set up an equation

$12 - 5x = -43$
$-12 \quad -12$
$-5x = -55$
$\div -5 \quad \div -5$
$x = 11$

OR We could start by adding $5x$ to both sides …
$12 - 5x = -43$
$+5x \quad +5x$
$12 = -43 + 5x$
and so on …

e.g. 4 $f(x) = 2x - 13$ and $g(x) = \frac{x+1}{3}$
Solve $f(x) = g(x)$

Use the information to set up an equation

$2x - 13 = \frac{x+1}{3}$
×3 ×3
$3(2x - 13) = 6x - 39 = x + 1$
$-x \quad -x$
$5x - 39 = 1$
$+39 \quad +39$
$5x = 40$
$\div 5 \quad \div 5$
$x = 8$

$\frac{x+1}{3} \times 3$ is NOT $3x + 3$

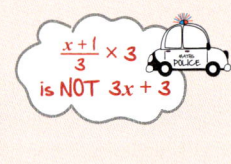

A. Given that $f(x) = 3x - 2$ find:

(i) $f(5)$ (ii) $f(12)$

(iii) $f(0)$ (iv) $f(-4)$

B. Given that $f(x) = x^2 + 7$ find:

(i) $f(3)$ (ii) $f(8)$

(iii) $f(-1)$ (iv) $f\left(\frac{3}{4}\right)$

C. Given that $g(x) = \frac{2x+1}{3}$ find:

(i) $g(13)$ (ii) $g(8.5)$

D. $f(x) = 2x - 9$
Find the value of x given that $f(x) = 21$

E. $g(x) = 6x + 1$
Solve $g(x) = 22$

F. $f(x) = \frac{4x - 7}{2}$
Find the value of x given that $f(x) = 12.5$

SET?

G. $f(x) = 4x + 1$
$g(x) = 55 - 2x$
$h(x) = \frac{3x + 40}{2}$

Solve:
(i) $f(x) = g(x)$
(ii) $g(x) = h(x)$
(iii) $f(x) = h(x)$

GO! FUNCTIONS 1

1. Place the digits 2, 3, 4 and 5 to complete the statement.

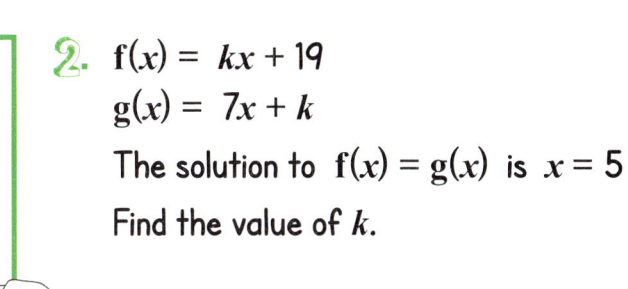

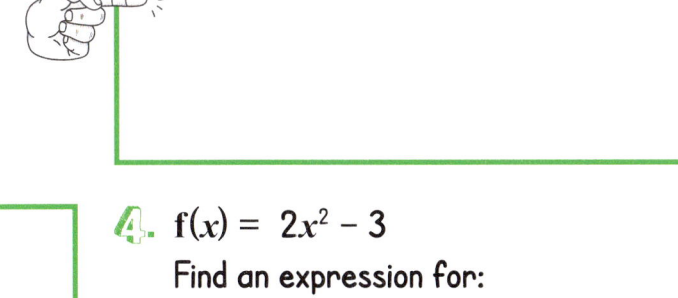

$f(x) = \square x^{\square} + \square$

$f(x) = 79$ when $x = \square$

2. $f(x) = kx + 19$
$g(x) = 7x + k$

The solution to $f(x) = g(x)$ is $x = 5$

Find the value of k.

3. Explain why $f(x) = \dfrac{1}{x}$ is not a function.

4. $f(x) = 2x^2 - 3$

Find an expression for:

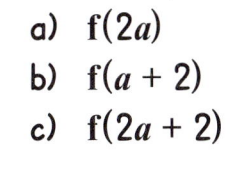

a) $f(2a)$
b) $f(a + 2)$
c) $f(2a + 2)$

READY?

We have seen that a function machine can be used to represent a function such as $f(x) = 3x + 5$

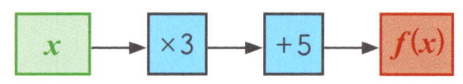

$x \to \times 3 \to +5 \to f(x)$

We can reverse a function machine to find a missing input. This is sometimes called an 'inverse'.

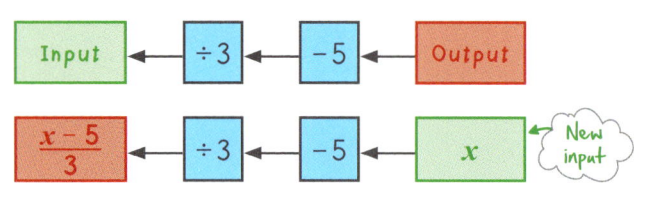

Input $\to \div 3 \to -5 \to$ Output

$\dfrac{x-5}{3} \leftarrow \div 3 \leftarrow -5 \leftarrow x$ (New input)

Using the notation for an **inverse function**, we write:

$$f^{-1}(x) = \dfrac{x-5}{3}$$

e.g. 1 Given that $f(x) = 6x - 7$ find $f^{-1}(x)$.

1 Write as a function machine:

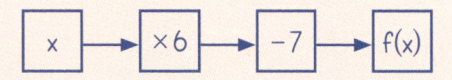

$x \to \times 6 \to -7 \to f(x)$

2 Work backwards, reversing the function machine:

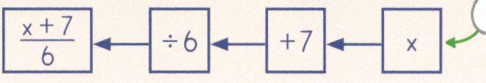

$\dfrac{x+7}{6} \leftarrow \div 6 \leftarrow +7 \leftarrow x$

The position of the input (x) has now changed

3 Write using the notation for an inverse function: $f^{-1}(x) = \dfrac{x+7}{6}$

OR We could rearrange the function:

$f(x) = 6x - 7$
$f(x) + 7 = 6x$
$\dfrac{f(x) + 7}{6} = x$
so ... $\dfrac{x+7}{6} = f^{-1}(x)$

CHECK-IN

1 Make x the subject:

$y = 6x$ \qquad $y = x - 45$

$y = 7x + 4$ \qquad $y = \dfrac{5(x-2)}{8}$

Functions can also be combined by using the output of the first as the input for the second, e.g.

$f(x) = 2x + 3$ and $g(x) = 4x - 10$

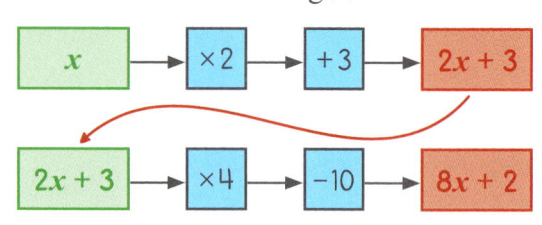

$x \to \times 2 \to +3 \to 2x+3$

$2x+3 \to \times 4 \to -10 \to 8x+2$

This **composite function** is written as $gf(x) = 8x + 2$

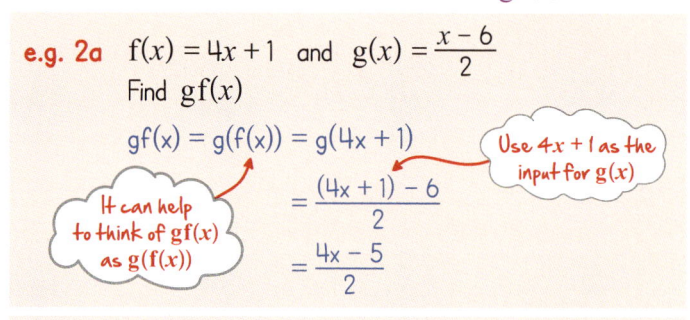

e.g. 2a $f(x) = 4x + 1$ and $g(x) = \dfrac{x-6}{2}$
Find $gf(x)$

$gf(x) = g(f(x)) = g(4x+1)$ *(Use $4x+1$ as the input for $g(x)$)*

$= \dfrac{(4x+1) - 6}{2}$

$= \dfrac{4x - 5}{2}$

It can help to think of $gf(x)$ as $g(f(x))$

e.g. 2b Find $fg(x)$. Give your answer in its simplest form.

$fg(x) = f(g(x)) = f\left(\dfrac{x-6}{2}\right)$

$4\left(\dfrac{x-6}{2}\right) = \dfrac{4(x-6)}{2} = 4\left(\dfrac{x-6}{2}\right) + 1$

$= \dfrac{4x - 24}{2} = 2x - 12 + 1$

$= 2x - 12 = 2x - 11$

$gf(x)$ is NOT the same as $fg(x)$

A. Find $f^{-1}(x)$ given that:

(i) $f(x) = 5x$

(ii) $f(x) = x + 6$

(iii) $f(x) = 2x - 17$

(iv) $f(x) = 7x + 2$

(v) $f(x) = 3(x - 2)$

B. Find $g^{-1}(x)$ given that:

(i) $g(x) = \dfrac{x}{7}$

(ii) $g(x) = \dfrac{x+13}{8}$

(iii) $g(x) = \dfrac{x-4}{15}$

(iv) $g(x) = \dfrac{8x+7}{2}$

SET?

C. Given that $f(x) = 4x - 3$
$g(x) = x^2 + 1$ and $h(x) = \dfrac{3x+1}{5}$
find:

(i) $fg(x)$

(ii) $hf(x)$

(iii) $gf(x)$

(iv) $hg(x)$

(v) $ff(x)$

GO! FUNCTIONS 2

Quality!

1. Di is asked to find $f^{-1}(x)$ if $f(x) = 10x + 1$

She writes:
$$y = 10x + 1$$
$$-1 \quad\quad -1$$
$$y - 1 = 10x$$
$$\div 10 \quad \div 10$$
$$\frac{y-1}{10} = x$$
$$\text{So } f^{-1}(x) = \frac{x-1}{10}$$

Do you agree with Di? Explain why.

2. $f(x) = \dfrac{5 - 2x}{6}$

Find $f^{-1}(x)$

3. $f(x) = \dfrac{5x - 1}{3}$

Show that $ff^{-1}(x) = x$

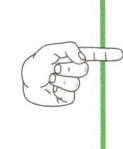

4. $f(x) = 6(x + 1)$ and $g(x) = \dfrac{2x + 7}{2}$

Solve $f^{-1}(x) = g^{-1}(x)$

5. $f(x) = (x + 2)^2$ and $g(x) = 3x + 4$

Show that:
$$3gf(x) - fg(x) = 12$$

READY?

The line with equation $y = \frac{2}{3}x - 1$ has a gradient of $\frac{2}{3}$ and a y-intercept of -1.

The line with equation $y = -\frac{3}{2}x + 3$ has a gradient of $-\frac{3}{2}$ and a y-intercept of 3.

If plotted on a coordinate grid, we see that these lines are **perpendicular**:

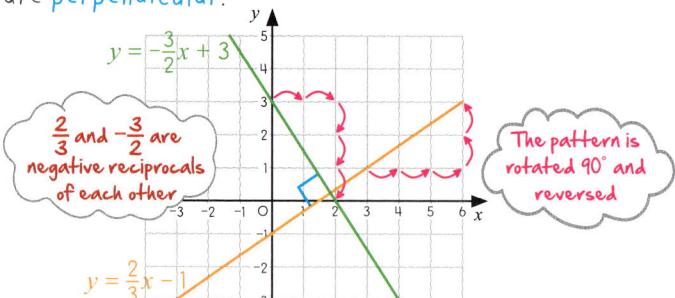

$\frac{2}{3}$ and $-\frac{3}{2}$ are negative reciprocals of each other

The pattern is rotated 90° and reversed

Any two lines that have gradients of $\frac{2}{3}$ and $-\frac{3}{2}$ are perpendicular.

The product of the gradients of perpendicular lines is -1

We can use our knowledge of reciprocals to help solve problems involving perpendicular lines …

e.g. 1 The gradient of line A is 4. Write down the gradient of the line that is perpendicular to A.

The negative reciprocal of a gradient m is $-\frac{1}{m}$: The gradient is $-\frac{1}{4}$

e.g. 2 The gradient of line B is $-\frac{3}{5}$. Work out the gradient of a line that is perpendicular to B.

The gradient is $-\frac{1}{-\frac{3}{5}} = -1 \times -\frac{5}{3} = \frac{5}{3}$

Notice that $-\frac{3}{5} \times \frac{5}{3} = -1$

CHECK-IN

1. Find the reciprocal:
   4 $\frac{1}{7}$ $-\frac{4}{9}$ 0.4

2. Make x the subject:
   $y = 7x + 2$ $y = \frac{x}{2} + 7$

3. Write down the gradient and y-intercept:
   $y = 5x + 8$ $y = 12 - \frac{1}{3}x$

e.g. 3 Here are the equations of six lines:

$y = 5x - 7$ $y = 3x + 5$ $y = \frac{1}{3}x - 7$

$y = \frac{1}{5}x + 5$ $y = -3x - 7$ $y = 3x$

Write down the equations of the two lines that are perpendicular.

1. Identify the gradients
2. Look for two gradients with a product of -1: $-3 \times \frac{1}{3} = -1$

so $y = -3x - 7$ and $y = \frac{1}{3}x - 7$ are perpendicular

e.g. 4 The graph shows line P with equation $y = 2x - 3$

Work out the equation of the line that passes though the point $(0, 5)$ and is perpendicular to P.

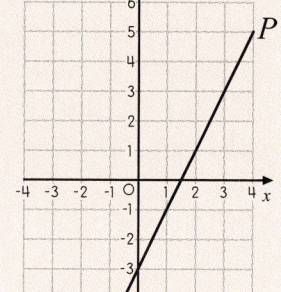

1. Find the gradient of the line:

The gradient of line P is 2, so the gradient of the perpendicular line is $-\frac{1}{2}$

2. Find the y-intercept of the line:

The y-intercept is 5 as the line passes through $(0, 5)$

3. Write the equation of the line:

$y = -\frac{1}{2}x + 5$

A. Work out the gradient of a line perpendicular to line A, given that the gradient of line A is:

(i) 6

(ii) -15

(iii) $-\frac{1}{5}$

(iv) $\frac{1}{8}$

(v) $\frac{2}{3}$

(vi) $-\frac{4}{9}$

B. Tick the functions whose graphs are perpendicular lines:

$y = -6x + 7$ $y = \frac{1}{5}x + 6$

$y = 5x + \frac{1}{6}$ $y = 6x - \frac{1}{6}$

$y = -\frac{1}{6}x - 7$ $y = -6x + 5$

C. Pair all the functions whose graphs are perpendicular lines:

$y = -\frac{1}{3}x + 4$ $y = \frac{1}{4}x + 3$

$y = 3 - 4x$ $y = 3x - \frac{1}{3}$

$y = -\frac{1}{4}x + 3$ $y = 4x + 3$

SET?

D. Line B has equation $y = \frac{1}{3}x + 4$

Line C is perpendicular to line B and passes through the point $(0, -8)$

Work out the equation of line C.

GO! PERPENDICULAR LINES

Genius!

1. Grace thinks that the lines with equations:
   $y = \frac{1}{5}x - 4$ and $2y = 6 - 5x$
   are perpendicular.
   Grace is not correct. Explain why.

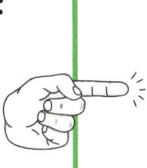

2. Here are the equations of six lines:

   $4y = x - 12$ $3y = 2x - 3$

   $x + 4y = 16$ $y = 3x - 5$

   $4x - y + 9 = 0$ $2y = 3x + 1$

   Write down the equations of the two lines that are perpendicular.

3. Line A passes through the points $(8,-2)$ and $(-1,-14)$.

   Line B is perpendicular to Line A and also passes through the point $(8,-2)$.

   Work out the equation of Line B.

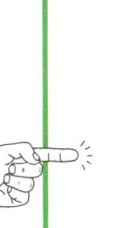

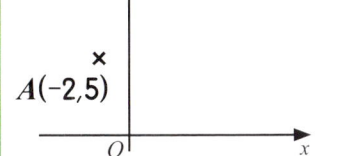

4. The point A has coordinates $(-2,5)$.
   The point B has coordinates $(4,15)$.
   The point M is the midpoint of A and B.

   Find the equation of the line that passes through M and is perpendicular to the line through A and B.

READY?

The quadratic expression $x^2 + 6x + 9$ is a perfect square as it can be written as $(x + 3)^2$ when factorised. Most quadratic expressions are not perfect squares. However, we can **complete the square** by making an adjustment. For example:

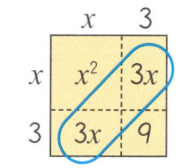

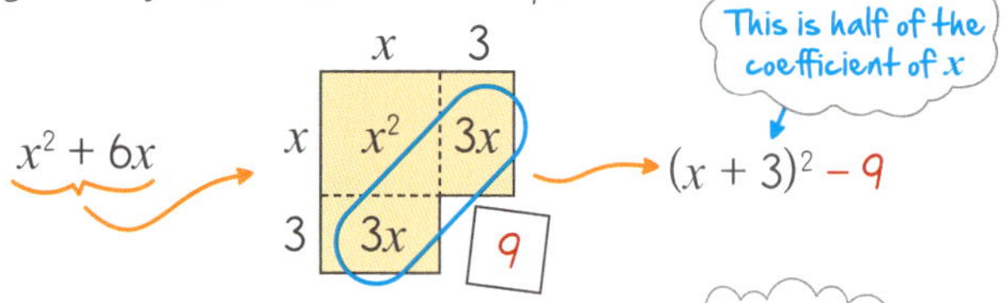

$x^2 + 6x \rightarrow (x + 3)^2 - 9$

$x^2 - 2x + 5 \rightarrow (x - 1)^2 - 1 + 5 = (x - 1)^2 + 4$

CHECK-IN

1. Factorise $x^2 + 8x + 16$
2. Factorise $x^2 + 6x + 8$
3. Factorise $9p^2 + 36p$
4. Expand and simplify $(x + 9)(x + 3)$
5. Calculate $\left(\frac{5}{2}\right)^2$
6. Work out $-\frac{7}{4} + 6$

e.g. 1
Write $x^2 + 8x$ in the form $(x + a)^2 + b$ where a and b are integers.

1 Halve the coefficient of x to find the value of a:
$8 \div 2 = 4$

2 Construct the square
$x^2 + 8x \xrightarrow{\div 2} (x + 4)^2$

3 Work out the adjustment to complete the square 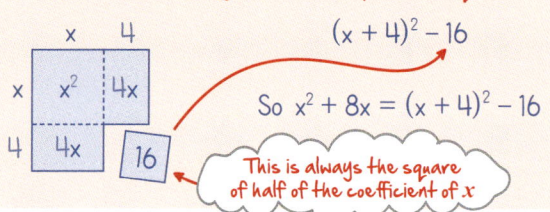
$(x + 4)^2 - 16$

So $x^2 + 8x = (x + 4)^2 - 16$

This is always the square of half of the coefficient of x

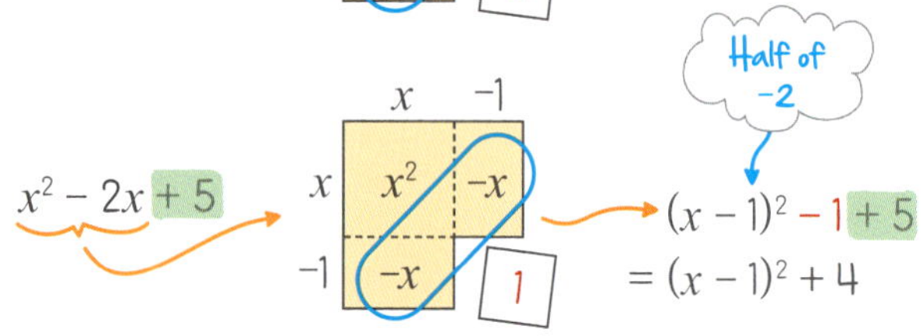

e.g. 2
Write $x^2 - 10x + 4$ in the form $(x + a)^2 + b$ where a and b are integers.

1 $-10 \div 2 = -5$

2 $x^2 - 10x \xrightarrow{\div 2} (x - 5)^2$

3 $x^2 - 10x = (x - 5)^2 - 25$

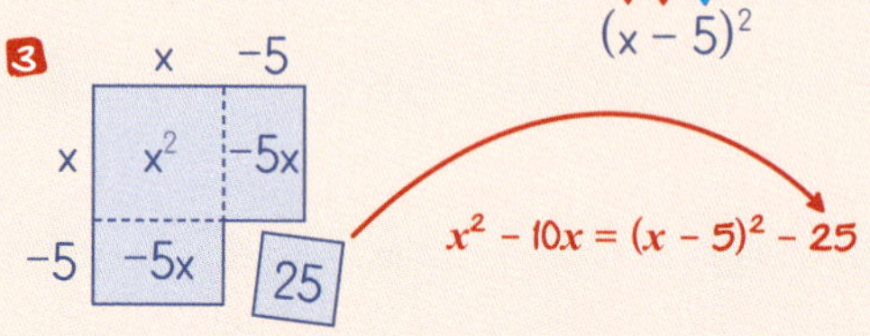

So $x^2 - 10x + 4 = (x - 5)^2 - 25 + 4$
$= (x - 5)^2 - 21$

TECHNICAL STUFF: $x^2 - 10x$ is equivalent to $x^2 + -10x$

e.g. 3
Write $x^2 + 3x - 7$ in the form $(x + p)^2 + q$

1 $3 \div 2 = \frac{3}{2}$

2 $x^2 + 3x \xrightarrow{\div 2} \left(x + \frac{3}{2}\right)^2$

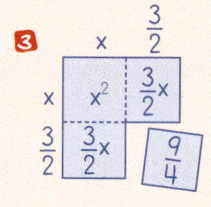

So $x^2 + 3x - 7 = \left(x + \frac{3}{2}\right)^2 - \frac{9}{4} - 7$
$= \left(x + \frac{3}{2}\right)^2 - \frac{37}{4}$

A. Write in the form $(x + a)^2 + b$

(i) $x^2 + 14x$

(ii) $x^2 - 20x$

(iii) $x^2 + 5x$

(iv) $x^2 - 9x$

B. Write in the form $(x + p)^2 + q$ where p and q are integers:

(i) $x^2 + 4x + 12$

(ii) $x^2 - 6x + 4$

(iii) $x^2 + 12x - 2$

(iv) $x^2 - 8x - 5$

SET?

C. Write in the form $(x + a)^2 + b$

(i) $x^2 + 7x - 1$

(ii) $x^2 - 5x + 3$

GO! COMPLETING THE SQUARE

1. Place the digits 1, 3, 6 and 8 to complete the statement.

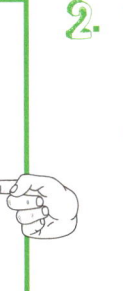

$$x^2 + \square x + \square = (x + \square)^2 - \square$$

2. Write
$$(x + 2)(x + 5)$$
in the form $(x + a)^2 + b$

3. By completing the square, solve:
$$x^2 - 18x + 3 = 0$$
Give your answers in the form $a \pm \sqrt{b}$

4. Write:
$$2x^2 - 12x + 5$$
in the form $p(x + q)^2 + r$ where p, q and r are integers.

HINT: Remove a factor of 2 first

Yay!

READY?

A **quadratic function** is one of the form $y = ax^2 + bx + c$. a, b and c can be any number. The graph of a quadratic function is always a **parabola**.

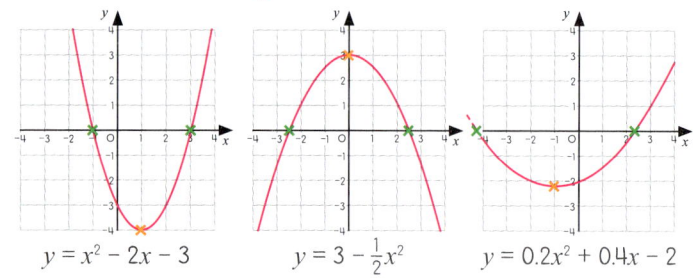

$y = x^2 - 2x - 3$ $y = 3 - \frac{1}{2}x^2$ $y = 0.2x^2 + 0.4x - 2$

A parabola has a line of symmetry through its **turning point**. It also identifies **roots** of related equations.

e.g. 1 Here is the graph of $y = -x^2 - 2x + 3$

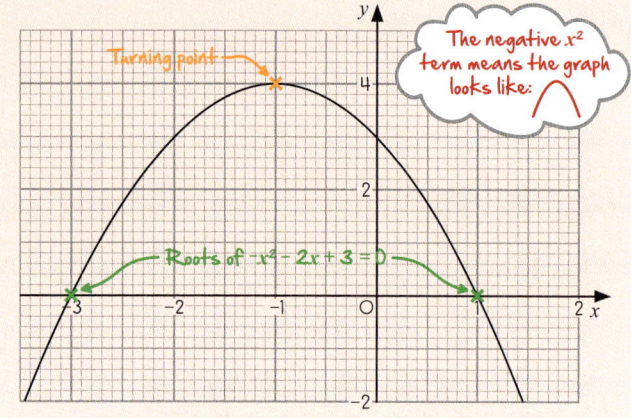

Turning point
The negative x^2 term means the graph looks like: ⌢
Roots of $-x^2 - 2x + 3 = 0$

a) Write down the turning point of the graph. $(-1, 4)$

b) Use the graph to find the roots of $x^2 + 2x = 3$

Compare the equation with $-x^2 - 2x + 3 = 0$

$x^2 + 2x = 3 \rightarrow -x^2 - 2x + 3 = 0$ so $x = -3$ and $x = 1$

CHECK-IN

1. Complete the table of values for $y = x^2 - 4x - 5$

x	-2	-1	0	1	2	3
y						

2. Find the solutions: $x^2 + 6x - 7 = 0$

3. Complete the square: $x^2 + 10x - 2$

4. Write down the roots and turning point:

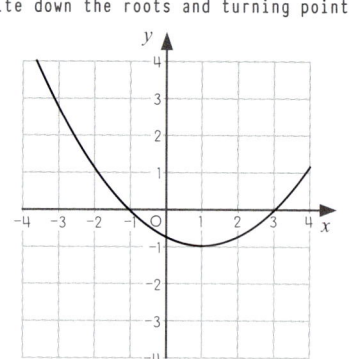

JUSTAROO Airlines
Ready for take off?

We can find roots and turning points without a graph...

e.g. 2 By completing the square, find the coordinates of the turning point of the graph of $y = x^2 + 5x + 9$

1 Complete the square for expression $x^2 + 5x + 9$

$5 \div 2 \rightarrow \frac{5}{2}$

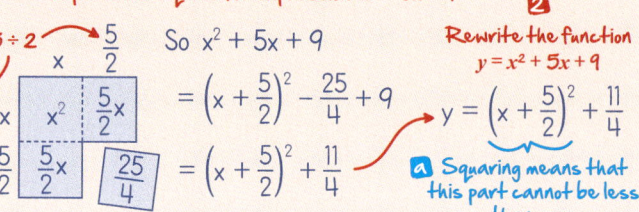

So $x^2 + 5x + 9$
$= \left(x + \frac{5}{2}\right)^2 - \frac{25}{4} + 9$
$= \left(x + \frac{5}{2}\right)^2 + \frac{11}{4}$

2 Rewrite the function
$y = x^2 + 5x + 9$
$y = \left(x + \frac{5}{2}\right)^2 + \frac{11}{4}$

a Squaring means that this part cannot be less than zero

3 Analyse the function:

b The positive x^2 term means the graph looks like: ⌣

c The smallest possible output of the function is when $x + \frac{5}{2} = 0$, so $x = -\frac{5}{2}$

when $x = -\frac{5}{2}$, $y = \frac{11}{4}$ so the turning point is $\left(-\frac{5}{2}, \frac{11}{4}\right)$

A. Here is the graph of $y = 5 + 8x - 2x^2$

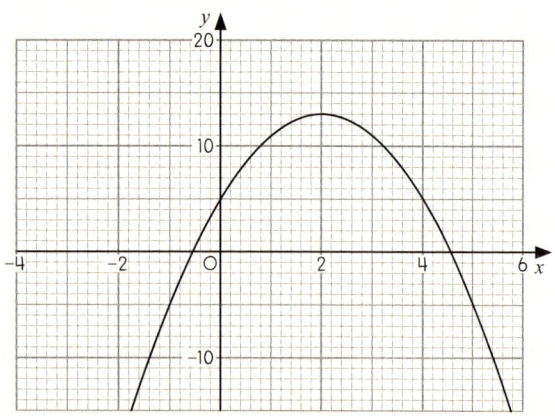

(i) Write down the turning point of the graph.

(ii) Use the graph to estimate the roots of the equation $2x^2 = 8x + 5$

B. (i) Write $x^2 + 14x - 5$ in the form $(x + a)^2 + b$ where a and b are integers:

(ii) Write down the turning point of the graph of $y = x^2 + 14x - 5$

SET?

C. By completing the square, find the coordinates of the turning point of the graph of $y = x^2 - 9x + 24$

46

GO! QUADRATIC FUNCTIONS

Great!

1. Josh is given the function $y = x^2 - 16x - 3$

He is asked to write the function in the form $y = (x + a)^2 + b$ and use this to find the coordinates of the turning point of the graph of $y = x^2 - 16x - 3$

Josh gives the coordinates of the turning point as $(-8, -67)$

Do you agree with Josh? Explain why.

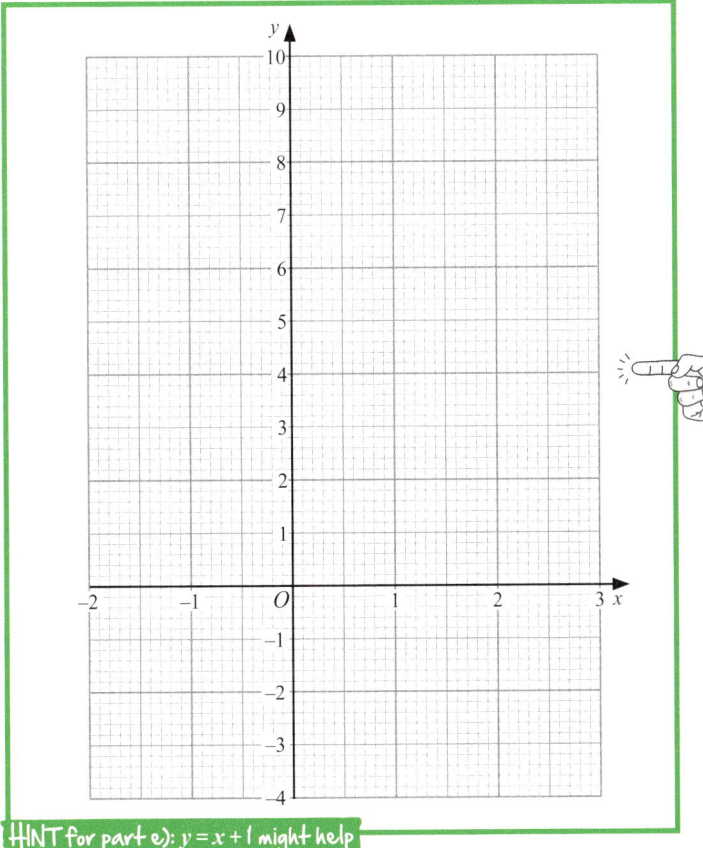

2. a) Complete the table of values for:
$$y = 8 + 3x - 3x^2$$

x	-1.5	-1	-0.5	0	0.5	1	1.5	2	2.5
y									

b) Plot the graph of $y = 8 + 3x - 3x^2$ on the grid.

c) Write down the turning point of the graph of $y = 8 + 3x - 3x^2$

d) Use the graph to estimate the roots of the equation:
$$3x^2 - 6 = 3x + 2$$

e) Use the graph to estimate the roots of the equation:
$$7 + 2x - 3x^2 = 0$$

HINT for part e): $y = x + 1$ might help

3. Here is an expression: $x^2 - 5x - 14$

a) Write the expression in the form $(x + a)^2 + b$

b) Find the turning point of the curve with equation $y = x^2 - 5x - 14$

c) Find the roots of $x^2 - 5x - 14 = 0$

d) Sketch the graph of $y = x^2 - 5x - 14$
Label the points of intersection with the axes.

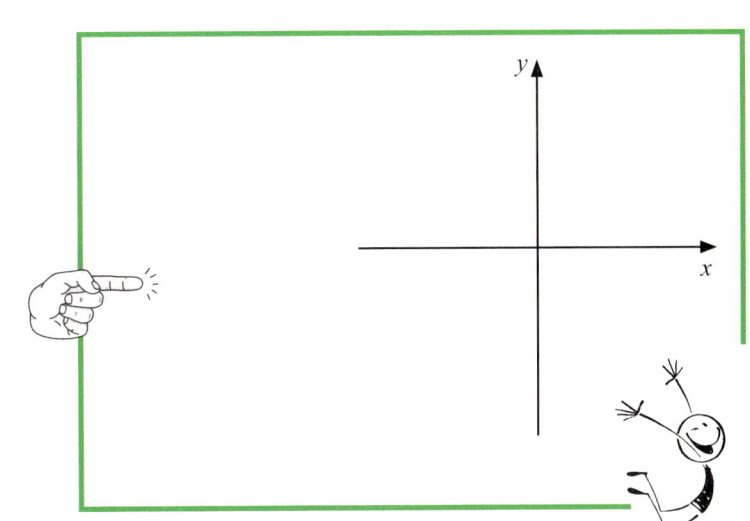

READY?

A **quadratic equation** is one of the form $ax^2 + bx + c = 0$ where a, b and c are any number. Quadratic equations often have two solutions. Our knowledge of factorising is helpful when solving quadratic equations:

CHECK-IN

Solve:
1. $(x - 7)(x - 2) = 0$
2. $y^2 + 6y = 0$
3. $x^2 + 4x - 32 = 0$
4. $x^2 + 15x + 36 = 0$

e.g. 1 Solve $2x^2 - 7x - 15 = 0$

1. Factorise the expression:
$2x^2 - 7x - 15 = (2x + 3)(x - 5)$

2. Rewrite the equation:
$(2x + 3)(x - 5) = 0$

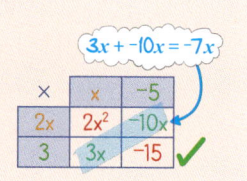

3. Solve the two possibilities:

$2x + 3 = 0$ and $x - 5 = 0$
$\quad -3 \quad -3 \qquad\qquad +5 \quad +5$
$2x = -3 \qquad\qquad x = 5$
$\div 2 \quad \div 2$
$x = -\dfrac{3}{2}$

4. Write the solutions:
$x = -\dfrac{3}{2}$ and $x = 5$

$3x + -10x = -7x$

×	x	-5
$2x$	$2x^2$	$-10x$
3	$3x$	-15

Sometimes rearrangement is needed first...

e.g. 3 Solve $3y^2 - y = 4$

$3y^2 - y = 4$
$-4 \qquad\qquad -4$
$3y^2 - y - 4 = 0$

We need the equation to be in the form $ay^2 + by + c = 0$

1. Factorise: $3y^2 - y - 4 = (3y - 4)(y + 1)$

2. Rewrite: $(3y - 4)(y + 1) = 0$

3. Solve: $3y - 4 = 0$ and $y + 1 = 0$
$\qquad\qquad 3y = 4 \qquad\qquad y = -1$
$\qquad\qquad y = \dfrac{4}{3}$

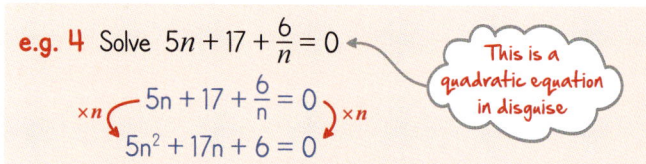

4. Write the solutions: $y = \dfrac{4}{3}$ and $y = -1$

e.g. 2 Solve $4x^2 - 25 = 0$

1. Use the difference of 2 squares to factorise: $a^2 - b^2 = (a + b)(a - b)$
$\sqrt{4x^2} = 2x$ and $\sqrt{25} = 5$ so $4x^2 - 25 = (2x + 5)(2x - 5)$

2. Rewrite: $(2x + 5)(2x - 5) = 0$

3. Solve: $2x + 5 = 0$ and $2x - 5 = 0$
$\qquad\quad -5 \quad -5 \qquad\qquad +5 \quad +5$
$\qquad\quad 2x = -5 \qquad\qquad 2x = 5$
$\qquad\quad \div 2 \quad \div 2 \qquad\qquad \div 2 \quad \div 2$
$\qquad\quad x = -\dfrac{5}{2} \qquad\qquad x = \dfrac{5}{2}$

4. Write the solutions: $x = -\dfrac{5}{2}$ and $x = \dfrac{5}{2}$

e.g. 4 Solve $5n + 17 + \dfrac{6}{n} = 0$

This is a quadratic equation in disguise

$\times n \quad 5n + 17 + \dfrac{6}{n} = 0 \quad \times n$
$5n^2 + 17n + 6 = 0$

1. Factorise: $5n^2 + 17n + 6 = (5n + 2)(n + 3)$

2. Rewrite: $(5n + 2)(n + 3) = 0$

3. Solve: $5n + 2 = 0$ and $n + 3 = 0$
$\qquad\quad 5n = -2 \qquad\qquad n = -3$
$\qquad\quad n = -\dfrac{2}{5}$

4. Write the solutions: $n = -\dfrac{2}{5}$ and $n = -3$

SET?

Solve:

A. $9x^2 - 4 = 0$

B. $81p^2 - 100 = 0$

C. $3x^2 - 19x - 14 = 0$

D. $5y^2 - 22y + 21 = 0$

E. $2x^2 + 15x = 8$

F. $7n^2 + 72n = -20$

G. $3x - 16 + \dfrac{5}{x} = 0$

H. $2k + \dfrac{12}{k} = 11$

48

GO! QUADRATIC EQUATIONS 1

Marvellous!

1. Solve $36 - 25x^2 = 0$

2. Solve $20 - 11a - 3a^2 = 0$

3. Solve $12y^2 + 54y + 60 = 0$

4. Gina is asked to solve the equation:
$$5x^2 + 17x = 12$$

She writes:
$$5x^2 + 17x = 12$$
$$5x^2 + 17x - 12 = 0$$
$$(5x - 3)(x + 4) = 0$$
so $x = -\frac{3}{5}$ and $x = 4$

Do you agree with Gina? Explain why.

The solutions to
$$\square x^2 - \square 7x + 30 = 0$$
are
$$x = \square \text{ and } x = \frac{\square}{7}$$

5. Place the digits 4, 5, 6 and 7 to complete the statement.

READY?

Some quadratic equations cannot be factorised, but we can always complete the square ...

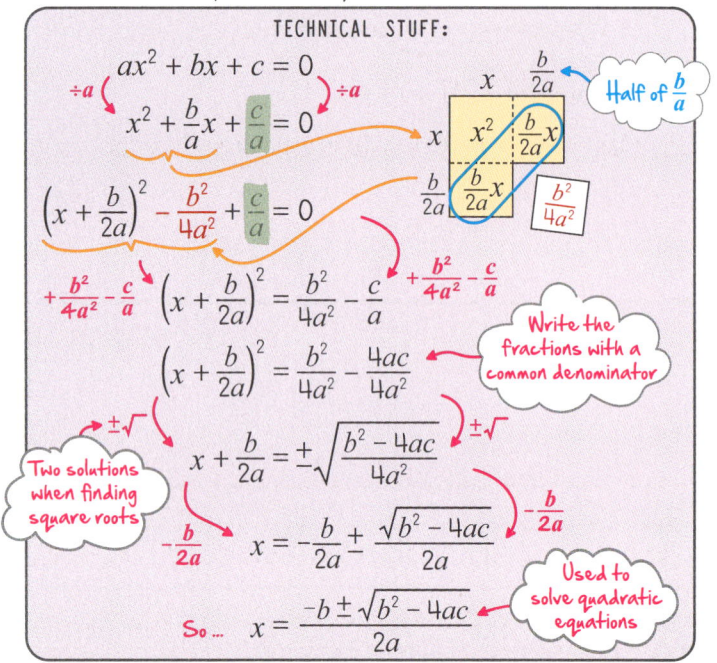

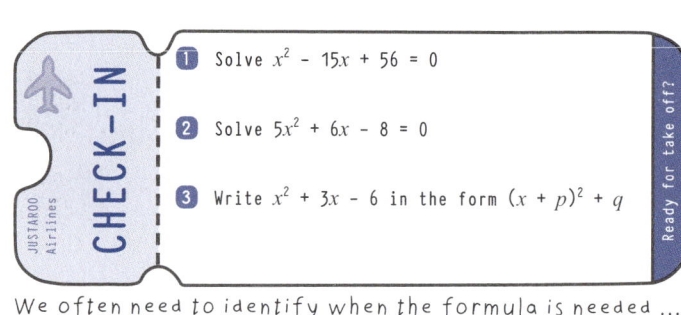

CHECK-IN

1. Solve $x^2 - 15x + 56 = 0$
2. Solve $5x^2 + 6x - 8 = 0$
3. Write $x^2 + 3x - 6$ in the form $(x + p)^2 + q$

TECHNICAL STUFF:

$$ax^2 + bx + c = 0$$
$\div a$ gives
$$x^2 + \frac{b}{a}x + \frac{c}{a} = 0$$

Half of $\frac{b}{a}$

$$\left(x + \frac{b}{2a}\right)^2 - \frac{b^2}{4a^2} + \frac{c}{a} = 0$$

$+\frac{b^2}{4a^2} - \frac{c}{a}$

$$\left(x + \frac{b}{2a}\right)^2 = \frac{b^2}{4a^2} - \frac{c}{a}$$

Write the fractions with a common denominator

$$\left(x + \frac{b}{2a}\right)^2 = \frac{b^2}{4a^2} - \frac{4ac}{4a^2}$$

$\pm\sqrt{}$

Two solutions when finding square roots

$$x + \frac{b}{2a} = \pm\sqrt{\frac{b^2 - 4ac}{4a^2}}$$

$-\frac{b}{2a}$

$$x = -\frac{b}{2a} \pm \frac{\sqrt{b^2 - 4ac}}{2a}$$

Used to solve quadratic equations

So ... $$x = \frac{-b \pm \sqrt{b^2 - 4ac}}{2a}$$

We often need to identify when the formula is needed ...

e.g. 1 Tick the equations which could be solved by factorising:

$3x^2 - 7x + 1 = 0$ ☐ $2y^2 + 11y + 14 = 0$ ☑
$x^2 + 5x - 24 = 0$ ☑ $x^2 - 3x - 12 = 0$ ☐

$x^2 + 5x - 24$

×	x	8
x	x^2	8x
−3	−3x	−24

$8x + -3x = 5x$ ✓

$= (x - 3)(x + 8)$

$2y^2 + 11y + 14$

×	2y	7
y	$2y^2$	7y
2	4y	14

$7y + 4y = 11y$ ✓

$= (y + 2)(2y + 7)$

The equations $3x^2 - 7x + 1 = 0$ and $x^2 - 3x - 12 = 0$ cannot be factorised ... the formula for solving quadratics could be used instead

To use the formula, we must identify the correct values:

e.g. 2 If $ax^2 + bx + c = 0$, state the value of a, b and c when $6x^2 + 7x - 12 = 0$

$a = 6$, $b = 7$ and $c = -12$

e.g. 3 Solve $5x^2 - 8x + 2 = 0$. Give your solutions to two significant figures.   **HINT:** Use the formula

1. Identify the values of a, b and c: $a = 5$, $b = -8$ and $c = 2$

   Negative -8 is the same as positive 8

2. Substitute into the formula: $x = \dfrac{-b \pm \sqrt{b^2 - 4ac}}{2a} = \dfrac{8 \pm \sqrt{(-8)^2 - 4 \times 5 \times 2}}{2 \times 5}$

   We might need FORMAT or S⇔D throughout

3. Use a calculator to evaluate each solution:

   ▭ 8 + √((−) 8)▢² − 4 × 5 × 2 ⌄ 2 × 5 = 1.2898...
   ▭ 8 − √((−) 8)▢² − 4 × 5 × 2 ⌄ 2 × 5 = 0.3101...

4. Round the solutions: $x = 1.3$ and $x = 0.31$ to two significant figures

---

A. Tick the equations which cannot be solved by factorising:

$2x^2 + x - 10 = 0$ ☐

$3k^2 - 11k + 9 = 0$ ☐

$4n^2 + 18n + 8 = 0$ ☐

$5x^2 - 9x - 7 = 0$ ☐

B. Given that $ax^2 + bx + c = 0$, state the value of a, b and c:

(i) $4x^2 + 7x + 2 = 0$
(ii) $4x^2 - 7x + 2 = 0$
(iii) $4x^2 + 7x - 2 = 0$
(iv) $-4x^2 - 7x + 2 = 0$

C. Solve $3x^2 + 5x + 1 = 0$

D. Solve $6x^2 + 23x - 5 = 0$

E. Solve $7x^2 - 12x + 4 = 0$

F. Solve $8x^2 - x - 3 = 0$

G. Solve $-3x^2 + 9x - 5 = 0$

Give all solutions to 2 decimal places

SET?

50

GO! QUADRATIC EQUATIONS 2

1. Solve $6x^2 - 17 = 3x$

Give your solutions to three significant figures.

2. Solve $5x + 1 = \dfrac{12}{x}$

Give your solutions to two decimal places.

3. Gabriel is asked to solve $5x^2 + 7x - 2 = 0$

He writes:

$a = 5, b = 7$ and $c = -2$

$x = -\dfrac{b \pm \sqrt{b^2 - 4ac}}{2a}$

$x = -\dfrac{7 \pm \sqrt{7^2 - 4 \times 5 \times -2}}{2 \times 5}$

so $x = 0.24...$ and $x = 1.64...$

Explain the mistake that Gabriel has made.

4. Here is a right-angled triangle:

- $(4x - 7)$ cm
- $(4x + 11)$ cm
- $3(2x + 1)$ cm

a) Work out the value of x
b) Find the length of the hypotenuse of the triangle.

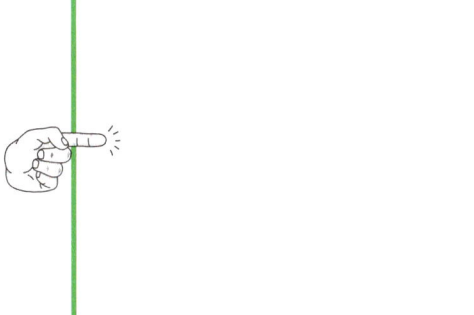

5. Solve $\dfrac{6x}{2x + 5} - \dfrac{3}{x - 7} = 0$

Give your solutions to three significant figures.

READY?

An expression such as 8^x has the variable in the power. Other names for 'power' include 'index' and 'exponent'. A function such as $y = 8^x$ is called an **exponential function**. Graphs of exponential functions always have a horizontal **asymptote**. This is a line that the curve approaches, but cannot cross.

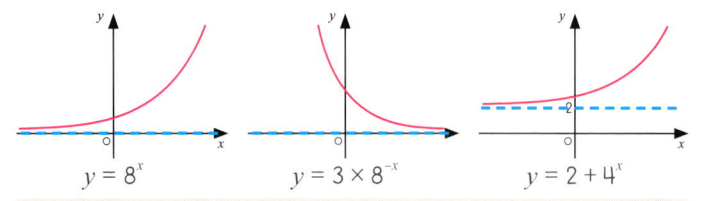

$y = 8^x$ $y = 3 \times 8^{-x}$ $y = 2 + 4^x$

CHECK-IN

1. Work out the value of y if $x = 4$:

   $y = 5^x$ $y = 2^{-x}$ $y = 10^{-x}$

   $y = (-3)^x$ $y = \left(\frac{1}{3}\right)^x$ $y = \left(-\frac{1}{10}\right)^x$

   $y = 60 \times 1.08^x$ $y = 200 \times \left(\frac{1}{2}\right)^x$

e.g. 1 a) Complete the table of values for $y = 3^x$

	A	B	C	D	E	F
x	-2	-1	0	1	2	3
y	$\frac{1}{9}$	$\frac{1}{3}$	1	3	9	27

When $x = -2$, $y = 3^{-2} = \frac{1}{3^2} = \frac{1}{9}$

When $x = 0$, $y = 3^0 = 1$

When $x = 3$, $y = 3^3 = 27$

b) On the grid below, plot the graph of $y = 3^x$

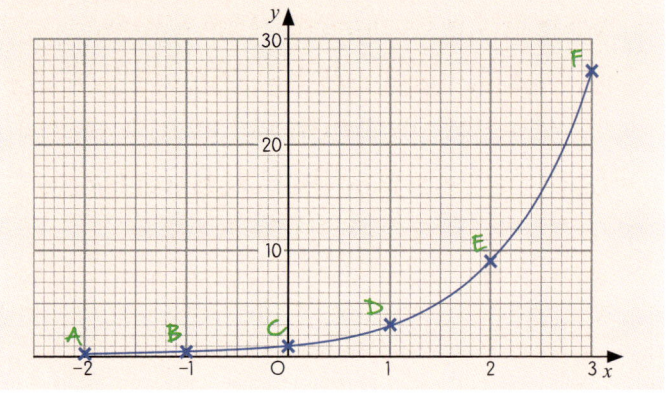

e.g. 2 a) Complete the table of values for $y = 25 + \frac{1}{2} \times 4^{-x}$

	A	B	C	D	E	F
x	-4	-3	-2	-1	0	1
y	153	57	33	27	25.5	25.125

b) Plot the graph of $y = 25 + \frac{1}{2} \times 4^{-x}$ on the grid.

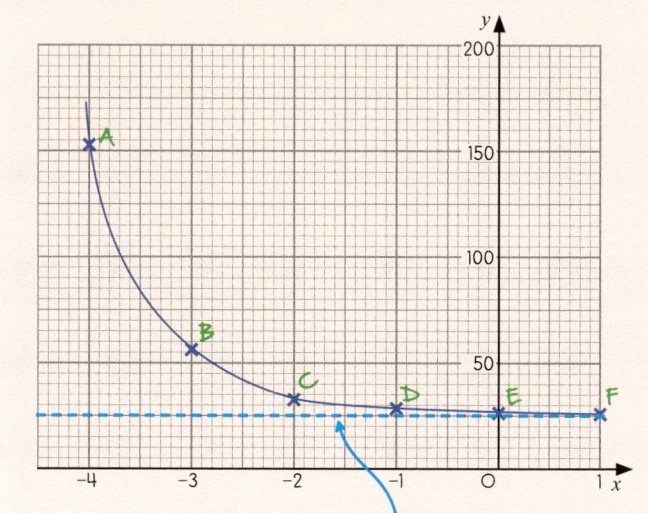

c) Write the equation of the asymptote of the graph

$y = 25$

Graphs of these functions show a picture of (1) exponential growth and (2) exponential decay

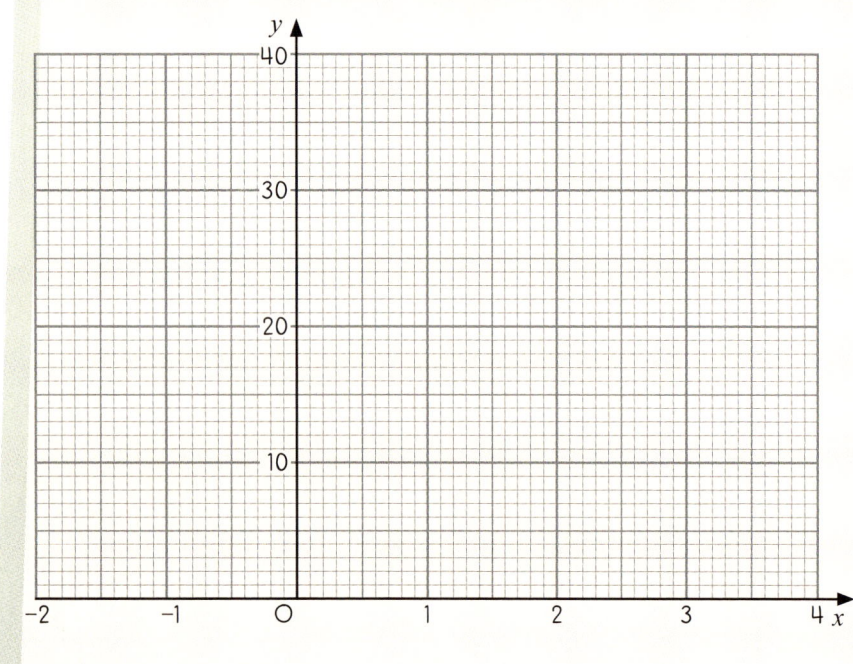

SET?

A. By completing the table of values, draw the graph of:

(i) $y = 2^x$

x	-2	-1	0	1	2	3	4
y							

(ii) $y = 5^{-x}$

x	-2	-1.5	-1	-0.5	0	0.5	1
y							

(iii) $y = 2.5 \times 2^x$

x	-2	-1	0	1	2	3	4
y							

EXPONENTIAL GRAPHS

1. The graphs of:

$y = 3^{-x}$
$y = 2^x$
$y = 6^x$
$y = x^2 + 1$

are plotted on the grid.

Write the equation of each graph in the correct box.

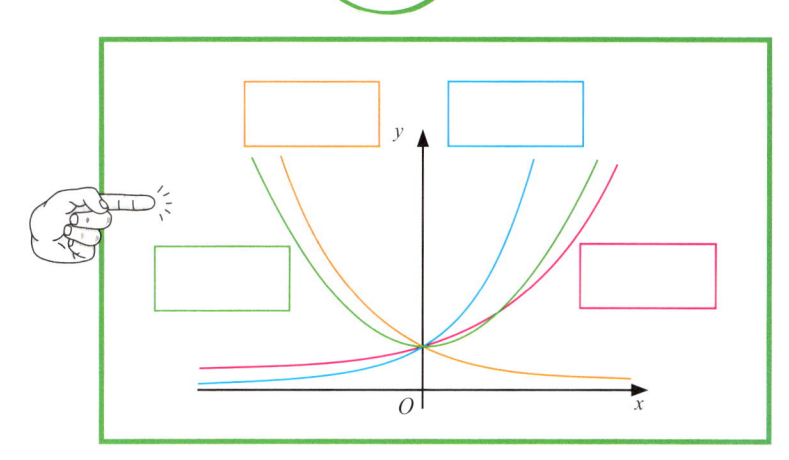

2. Greg invests £500 in a long-term savings account offering 4.5% interest per year.

The function $V = 500 \times 1.045^t$ gives the value of the investment after t years.

a) Plot the graph of $V = 500 \times 1.045^t$ for values of the investment up to 5 years.

b) Greg thinks that the function is linear. Explain why this is not the case.

3. A radioactive substance decays over time.

The formula $M = 1000 \times \left(\frac{1}{2}\right)^{\frac{n}{3}}$ can be used to work out the mass in grams remaining after n years.

a) What does the number 1000 represent?

b) Plot the graph of $M = 1000 \times \left(\frac{1}{2}\right)^{\frac{n}{3}}$ for $0 \leq n \leq 5$

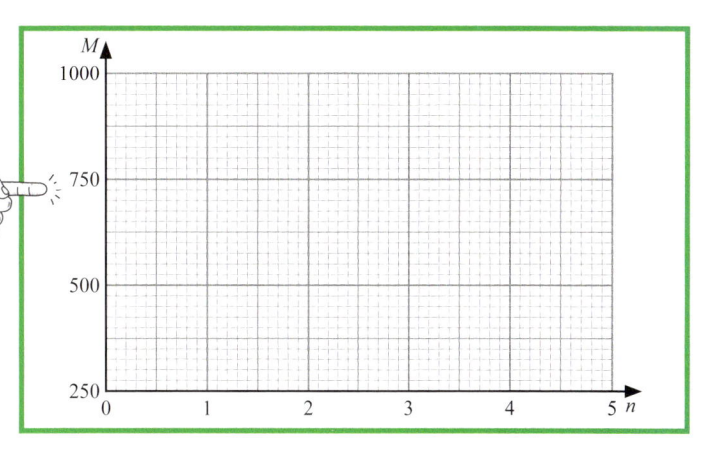

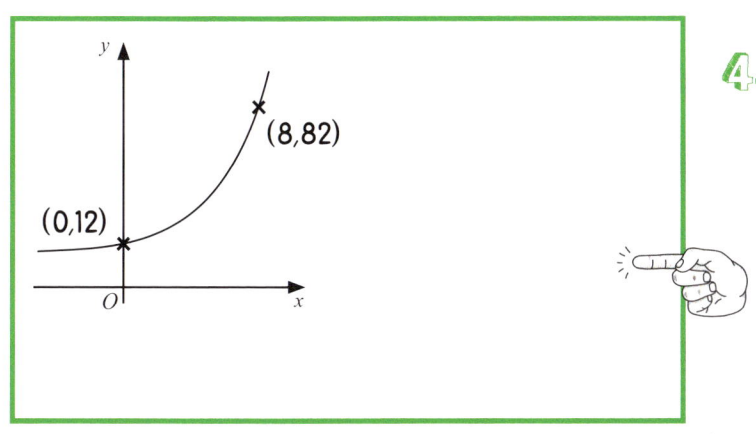

4. A graph has equation $y = A + B \times 6^{\frac{x}{4}}$

The curve passes through the points with coordinates (0,12) and (8,82).

Work out the values of A and B.

READY?

Trigonometry is not only connected to angles in a triangle. Angles of any size can be used as inputs for a trigonometric function. For example:

$\sin 150° = 0.5 \qquad \cos 270° = 0 \qquad \tan 495° = -1$

Sets of angles have the same output. For example:

$\sin 30° = \sin 150° = \sin 390° = \sin 510°$ etc. $= 0.5$

We can plot a graph of the function to see the pattern:

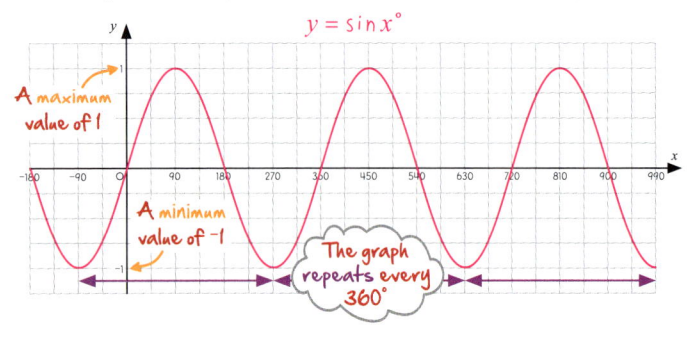

$y = \sin x°$
- A maximum value of 1
- A minimum value of -1
- The graph repeats every 360°

The graph of $y = \cos x°$ is similar. It also has a minimum of -1, a maximum of 1, and repeats every 360°.

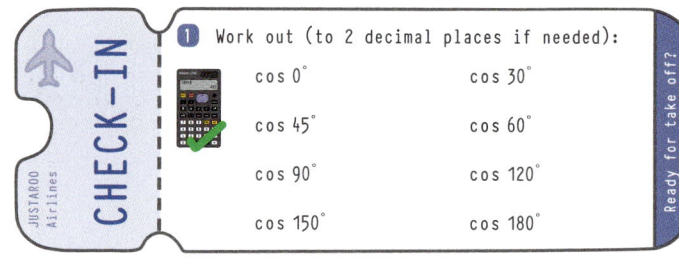

The graph of $y = \tan x°$ is different...

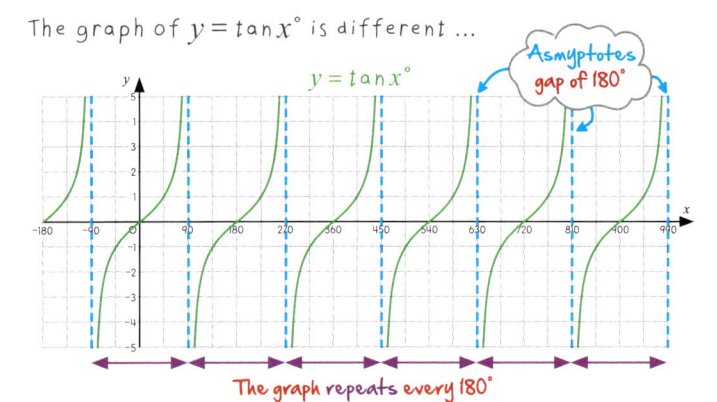

$y = \tan x°$
- Asymptotes gap of 180°
- The graph repeats every 180°

CHECK-IN

1 Work out (to 2 decimal places if needed):

$\cos 0° \qquad \cos 30°$

$\cos 45° \qquad \cos 60°$

$\cos 90° \qquad \cos 120°$

$\cos 150° \qquad \cos 180°$

We can solve problems using trigonometric graphs...

e.g. Here is the graph of $y = \tan x°$ for values of x between 0 and 300

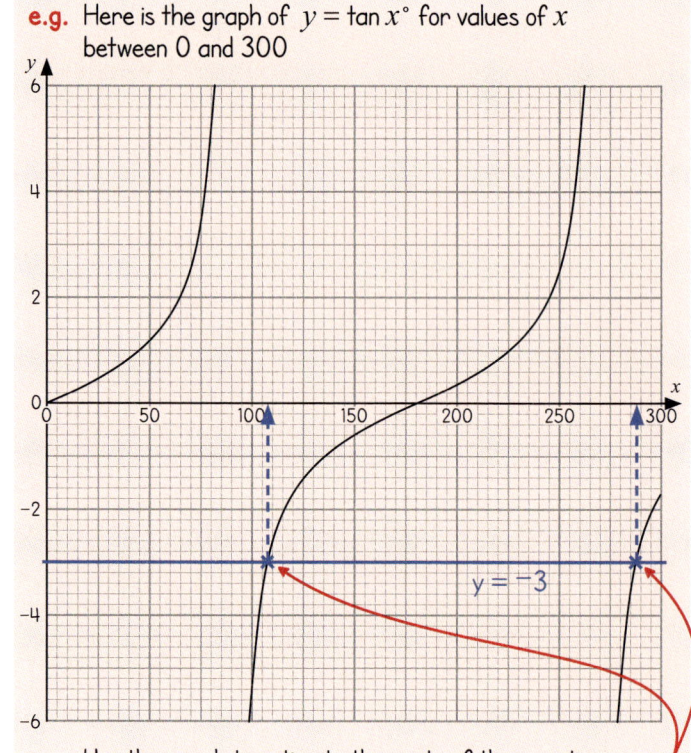

Use the graph to estimate the roots of the equation $4 \tan x = -12$ where x is an angle between 0° and 300°

1 Make $\tan x$ the subject:

$4 \tan x = -12$
$\div 4 \qquad \div 4$
$\tan x = -3$

2 Draw the line $y = -3$ and look for points of intersection...

3 Read the values off the x-axis:

$x = 108°$ and $x = 288°$

SET?

A. Here is the graph of $y = \sin x°$ for values of x between 0 and 180

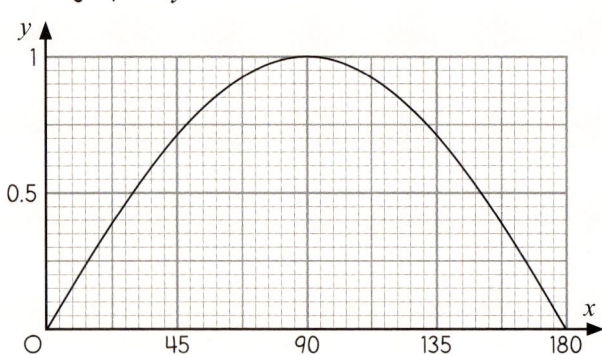

Use the graph to estimate the solutions, between 0 and 180, of:

(i) $\sin x = 0.25$

(ii) $20 \sin x = 13$

B. Sketch the graph of $y = \cos x°$ for values of x between 0 and 360

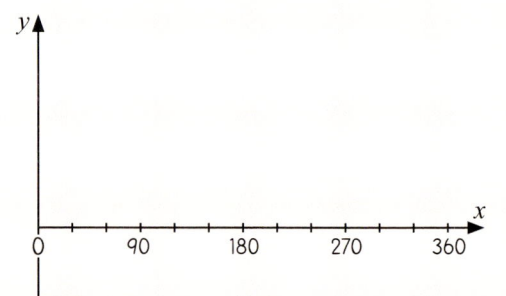

GO! TRIGONOMETRIC GRAPHS

slay!

1. Phil is asked to sketch the graph of the function $y = \tan x°$ for values of x between 0 and 360. Here is his sketch:

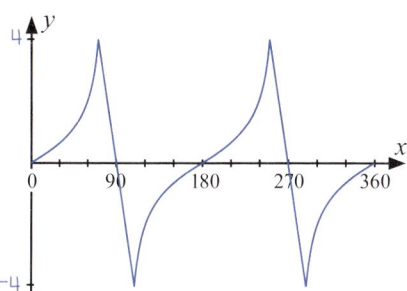

Explain why Phil is not correct.

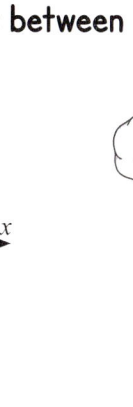

2. Sea level in the UK is measured from a point in Newlyn harbour, Cornwall. The depth of water, in metres, in the harbour can be modelled by the function:

$$D = 3.5 - 2\sin(30x° - 100)$$

where x is the number of hours after midnight on 30th June 2024

Plot the graph of the function for values of x between 0 and 10

3. a) The exact value of $\tan 60°$ is $\sqrt{3}$
Explain why $\tan 600° = \sqrt{3}$

b) The exact value of $\sin 15°$ is $\dfrac{\sqrt{6} - \sqrt{2}}{4}$

Explain why $\sin 165° = \dfrac{\sqrt{6} - \sqrt{2}}{4}$

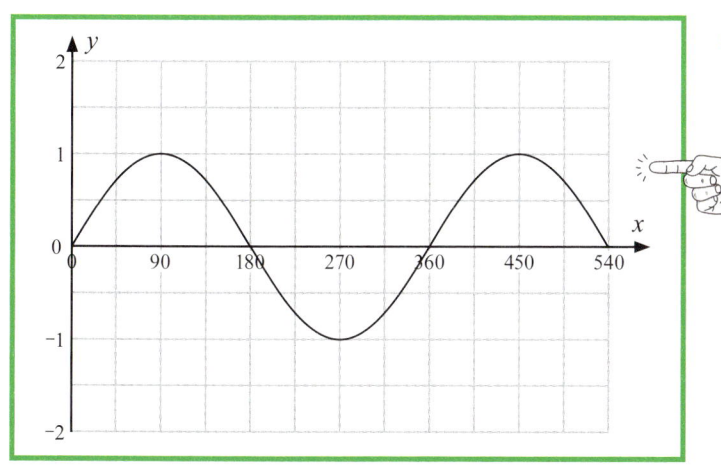

4. Here is part of the graph of $y = \sin x°$

Use the graph to work out the number of solutions to the equation:

$$\sin x° = 2 - \frac{x}{120}$$

Tick the correct number of solutions:

☐ 0   ☐ 1

☐ 2   ☐ More than 2

READY?

If we compare the graph of $y = x^3 + 2x^2$ with the graph of $y = x^3 + 2x^2 + 2$, we see the same curve but in a different place:

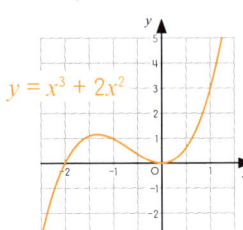

$y = x^3 + 2x^2$

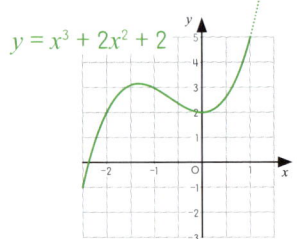

$y = x^3 + 2x^2 + 2$

The second curve is 2 units higher. In general, $y = f(x) + a$ translates $y = f(x)$ vertically by a units. We can also say that ...

$y = f(x) + a$ translates the graph $y = f(x)$ by a vector $\begin{pmatrix} 0 \\ a \end{pmatrix}$

If we compare the graphs of $y = x^2$ and $y = (x + 3)^2$, we see the same curve in a different place again:

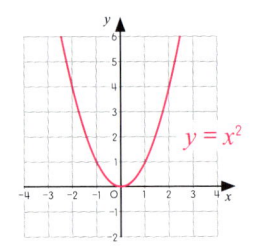

$y = x^2$

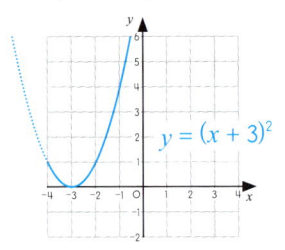

$y = (x + 3)^2$

This time the second curve is 3 units to the left. In general, $y = f(x + a)$ translates $y = f(x)$ horizontally by $-a$ units. We can also say that ...

$y = f(x + a)$ translates the graph $y = f(x)$ by a vector $\begin{pmatrix} -a \\ 0 \end{pmatrix}$

CHECK-IN

1 Translate triangle A using the vector $\begin{pmatrix} -2 \\ 0 \end{pmatrix}$

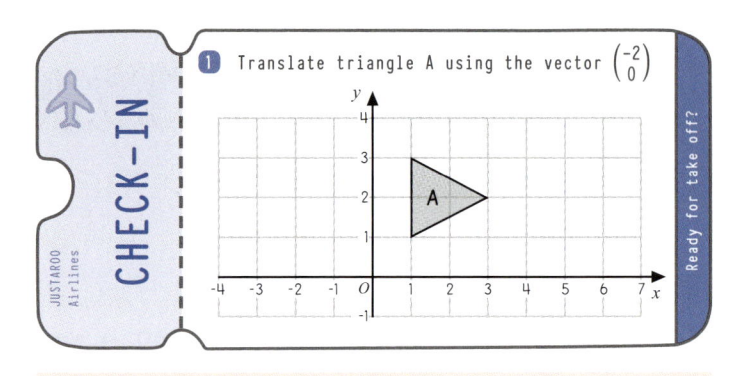

e.g. 1 The graph of $y = f(x)$ is shown on the grid.

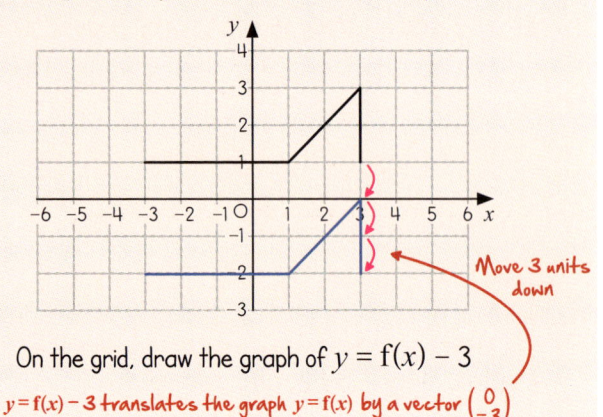

On the grid, draw the graph of $y = f(x) - 3$

Move 3 units down

$y = f(x) - 3$ translates the graph $y = f(x)$ by a vector $\begin{pmatrix} 0 \\ -3 \end{pmatrix}$

e.g. 2 The graph of $y = f(x)$ is shown on the grid.

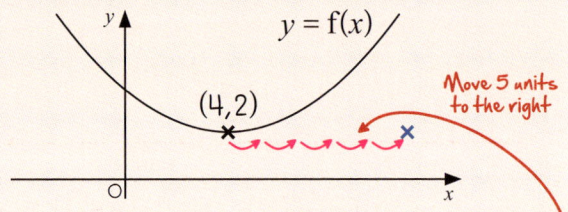

Move 5 units to the right

The turning point of the curve is at $(4, 2)$
Find the coordinates of the turning point of the graph of $y = f(x - 5)$

This translates the graph by a vector $\begin{pmatrix} 5 \\ 0 \end{pmatrix}$

The turning point will be at $(9, 2)$

A. Here is the graph of $y = f(x)$

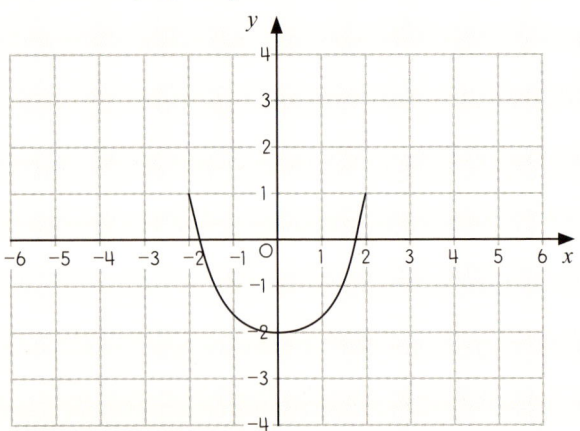

On the grid, sketch and label the graph of:

(i) $y = f(x + 4)$

(ii) $y = f(x) - 2$

B. The graph of $y = f(x)$ is shown on the grid.

$P(0, 4)$

$Q(4, 2)$

$y = f(x)$

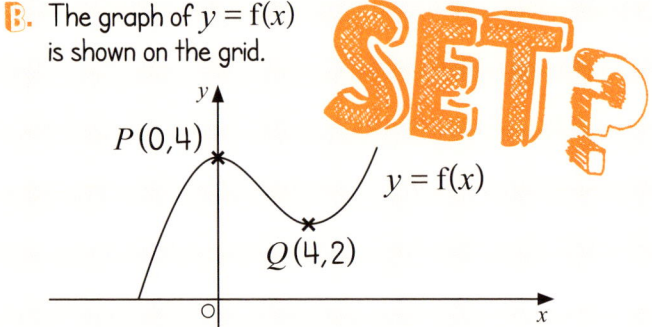

P and Q are turning points on the curve.

(i) Find the coordinates of P on the graph of
- $y = f(x + 6)$
- $y = f(x) - 12$

(ii) Find the coordinates of Q on the graph of
- $y = f(x - 8)$
- $y = f(x) + 15$

SET?

56

GO! TRANSFORMING GRAPHS 1

Belter!

1. Bev is given the graph of $y = f(x)$ and asked to sketch the graph of $y = f(x + 4)$ on the same grid. Here is her sketch:

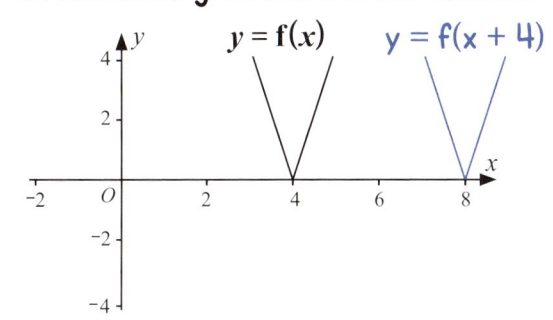

Explain why Bev is not correct.

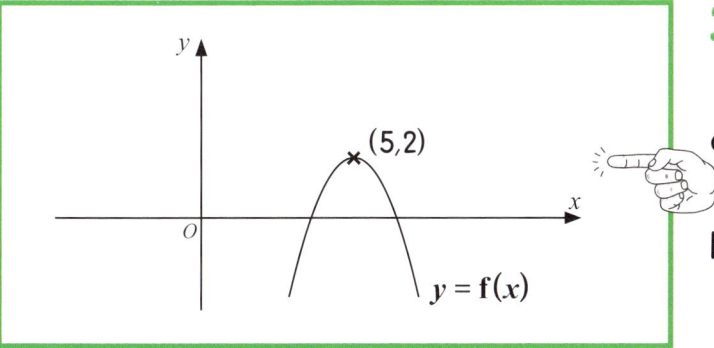

2. The graph of $y = f(x)$ is shown on the grid. The curve has a turning point at $(5,2)$

   a) On the same grid, sketch the graph of:
   $$y = f(x + 6) + 1$$
   b) Label the coordinates of the turning point on your graph.

3. Here is the graph of $y = f(x)$

   The graph has two asymptotes.

   State the equation of the two asymptotes on the graph of $y = f(x) - 15$

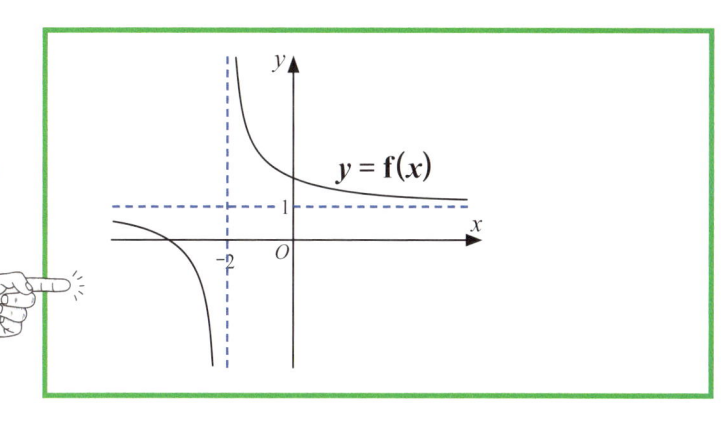

4. Here is part of the graph of $y = \cos x°$

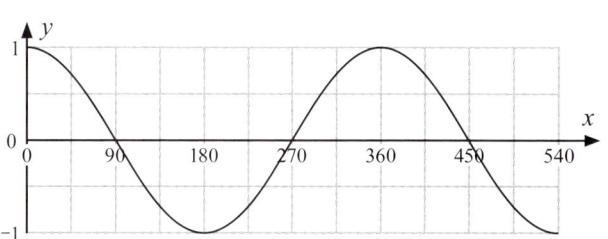

   Tick the statements that are always true.

   ☐ $\cos x° = \cos (x + 180)°$

   ☐ $\sin x° = \cos (x - 90)°$

   ☐ $\cos x° = \sin (x - 90)°$

   ☐ $\cos x° = \sin (x + 90)°$

READY?

Plotting the graphs of $y = (x + 4)^3$ and $y = (-x + 4)^3$ on the same grid shows that the y-axis is a line of symmetry between them:

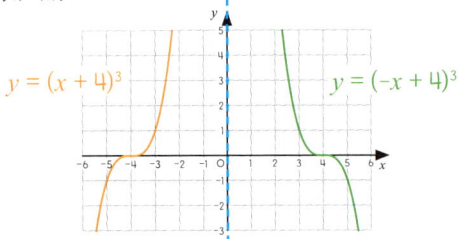

In general, the graph of $y = f(-x)$ is a reflection of the graph of $y = f(x)$ in the y-axis. We could also say ...

$y = f(-x)$ is a reflection of $y = f(x)$ in the line $x = 0$

Sometimes the x-axis is a line of symmetry. We can see an example of this by plotting the graphs of $y = x^2 + 1$ and $y = -(x^2 + 1)$:

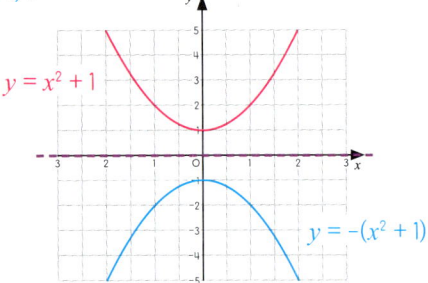

In general, the graph of $y = -f(x)$ is a reflection of the graph of $y = f(x)$ in the x-axis. We could also say ...

$y = -f(x)$ is a reflection of $y = f(x)$ in the line $y = 0$

CHECK-IN

1 Reflect the rectangle in the line $x = 0$

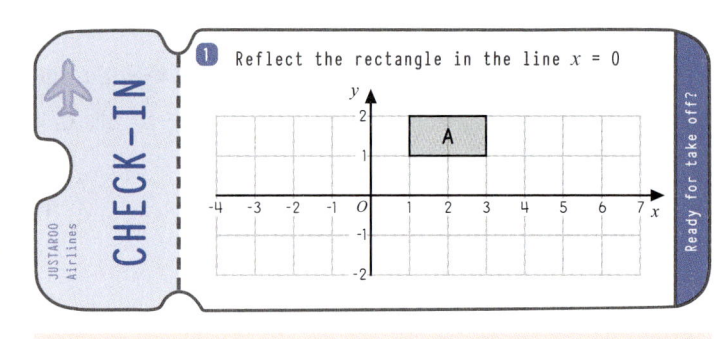

e.g. 1 The graph of $y = f(x)$ is shown on the grid.

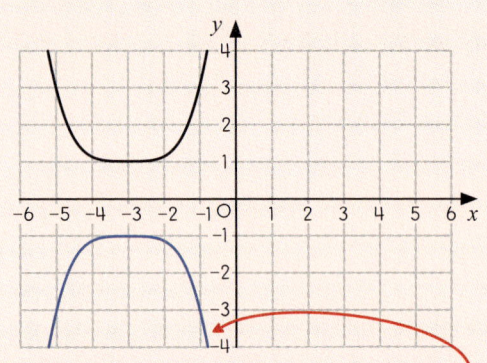

On the grid, draw the graph of $y = -f(x)$

$y = -f(x)$ reflects the graph in the x-axis

e.g. 2 The graph of $y = f(x)$ is shown on the grid.

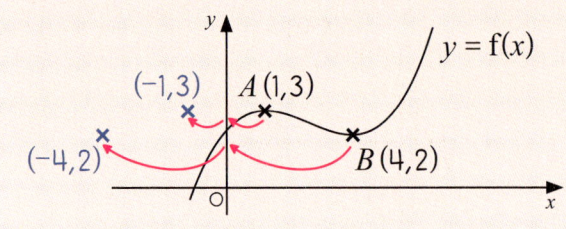

A and B are turning points of the curve.
Find the coordinates of the turning points of the graph of $y = f(-x)$

$y = f(-x)$ reflects the graph in the y-axis
The turning points will be at $(-1, 3)$ and $(-4, 2)$

SET?

A. Here is the graph of $y = f(x)$

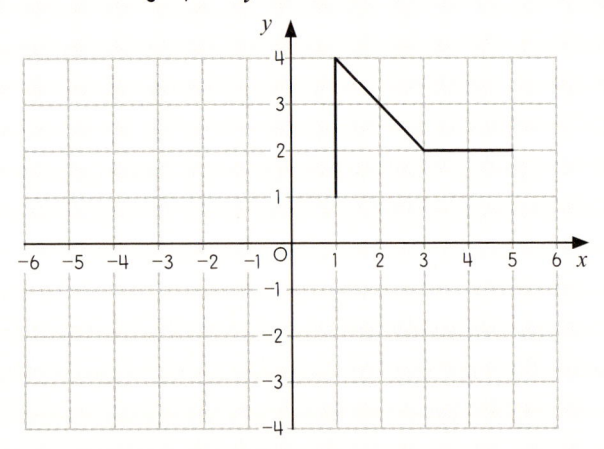

On the grid, sketch and label the graph of:
(i) $y = f(-x)$
(ii) $y = -f(x)$

B. The graph of $y = f(x)$ is shown on the grid.

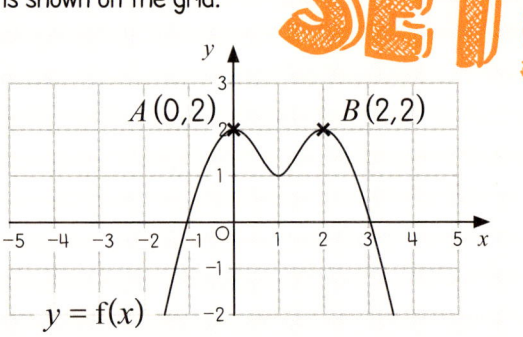

A and B are turning points on the curve.
Complete the table for the coordinates of A of B on the graphs of:
(i) $y = f(-x)$
(ii) $y = -f(x)$

	A	B
$y = f(-x)$		
$y = -f(x)$		

58

GO! TRANSFORMING GRAPHS 2

1. The graph of $y = f(x)$ is shown on the grid.

On the grid, draw and label the graph of:
$$y = f(-x) - 3$$

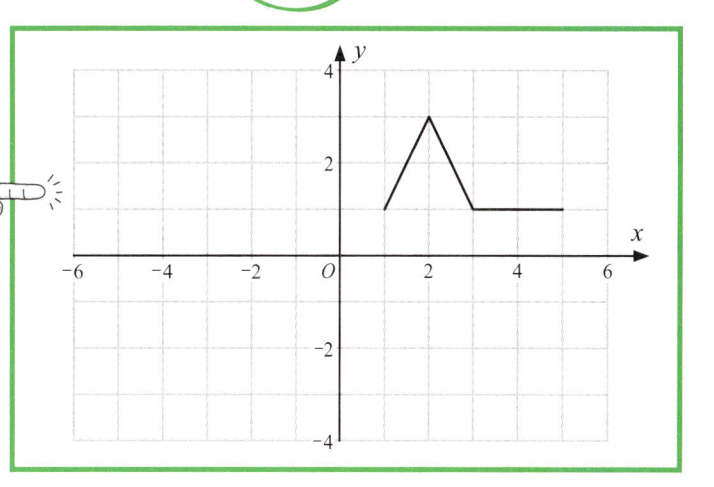

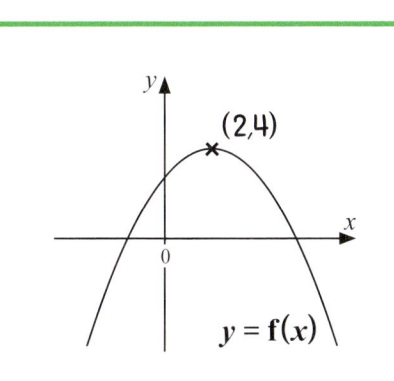

2. The graph of $y = f(x)$ is shown on the grid.

The curve has a turning point at $(2,4)$

Find the coordinates of the turning point of:

a) $y = -f(x) + 3$
b) $y = f(-x) + 3$
c) $y = -f(x + 3)$

3. Here is the graph of $y = f(x)$.

The curve has a turning point at $(-3,2)$

The graph of $y = -f(x) + k$ also has a turning point at $(-3,2)$

State the value of k.

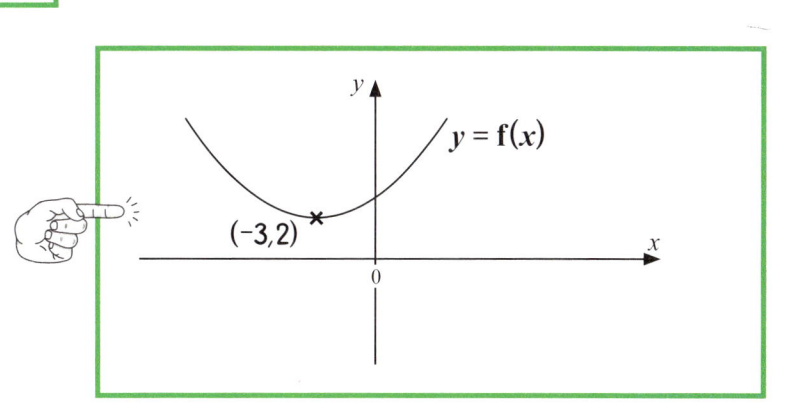

$y = x^3$ ☐

$y = \cos x°$ ☐

$y = \tan x°$ ☐

$y = \sin x°$ ☐

$y = x^2$ ☐

4. If the graph of $y = f(x)$ is the same as the graph of $y = f(-x)$, then $f(x)$ is an even function.

If the graph of $y = -f(x)$ is the same as the graph of $y = f(-x)$, then $f(x)$ is an odd function.

State whether each of these functions is even or odd.

READY?

The **gradient** of a straight line is a measure of its steepness. For every one unit to the right, the gradient states the number of units up*. When the equation of a line is written as $y = mx + c$, the gradient appears as 'm' [from the French, 'monter': to climb].

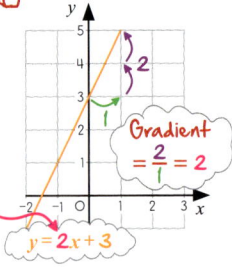

Gradient = $\frac{2}{1}$ = 2

$y = 2x + 3$

The gradient of a curve is constantly changing. At every point on the curve, the steepness is measured by the gradient of the tangent at that point. For example ...

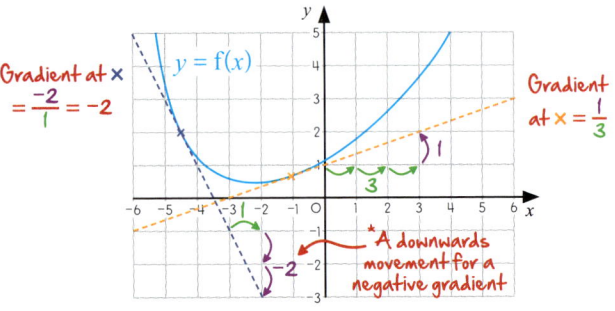

Gradient at × = $\frac{-2}{1}$ = -2

Gradient at x = $\frac{1}{3}$

*A downwards movement for a negative gradient

We often need to estimate the gradient of a curve at a point. This involves drawing a tangent as accurately as possible ...

e.g. 1 Draw a tangent to the curve at the point (0.5, 1.5)

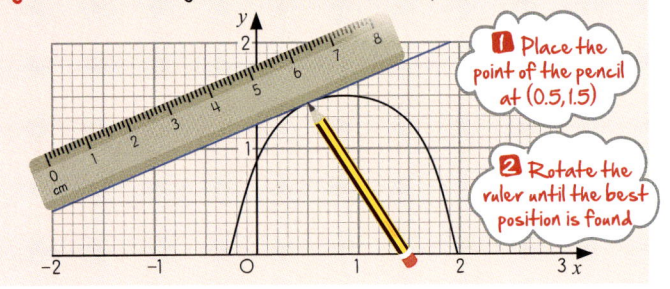

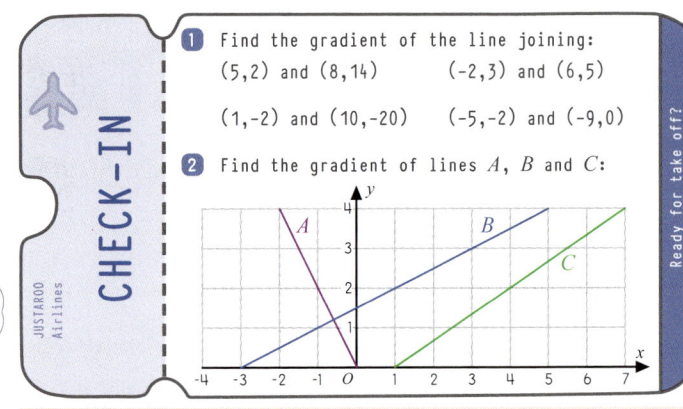

1. Place the point of the pencil at (0.5, 1.5)
2. Rotate the ruler until the best position is found

CHECK-IN

1. Find the gradient of the line joining:
   (5, 2) and (8, 14)   (-2, 3) and (6, 5)
   (1, -2) and (10, -20)   (-5, -2) and (-9, 0)

2. Find the gradient of lines A, B and C:

e.g. 2 The graph of $y = f(x)$ is shown on the grid.

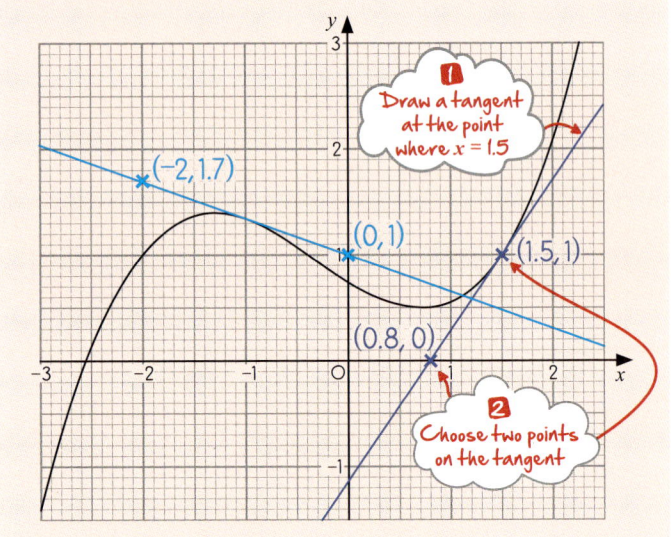

1. Draw a tangent at the point where $x = 1.5$
2. Choose two points on the tangent

a) Estimate the gradient of the graph when $x = 1.5$

3. Work out the gradient using $\frac{\text{change in } y}{\text{change in } x}$

$$\frac{1 - 0}{1.5 - 0.8} = \frac{1}{0.7} = 1.4285... = 1.43 \text{ to 2 decimal places}$$

b) Estimate the gradient of the graph when $x = -1$

$$\frac{1 - 1.7}{0 - -2} = \frac{-0.7}{2} = -0.35$$

The gradient here is NOT positive

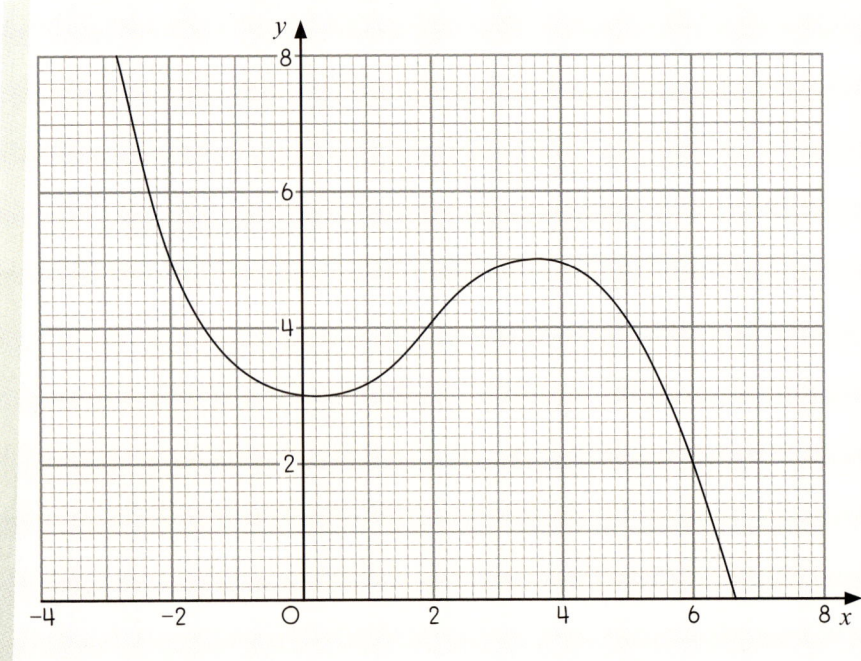

SET?

A. Here is the graph of $y = f(x)$

(i) Estimate the gradient of the graph at the point (6, 2)

(ii) Estimate the gradient of the graph when $x = 1$

(iii) Estimate the gradient of the graph at the point (-2, 5)

GO! ESTIMATING GRADIENTS

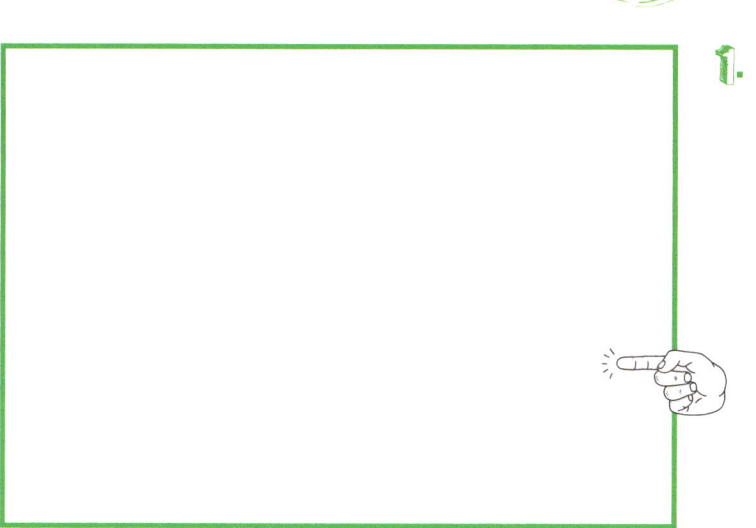

stunning!

1. Here is the graph of $y = \frac{1}{2}x^2 - 2$

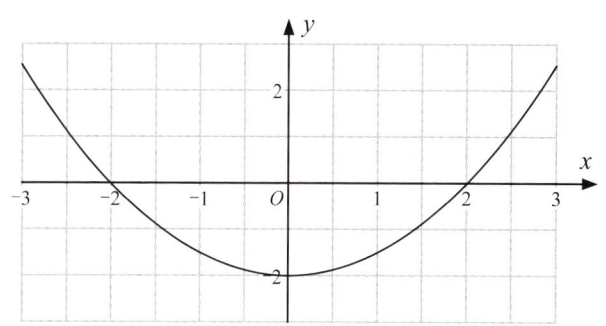

   Kate says: "The gradient of the graph is $\frac{1}{2}$."
   Explain why Kate is incorrect.

2. Here is the graph of $y = x^2 - 7x + 11$

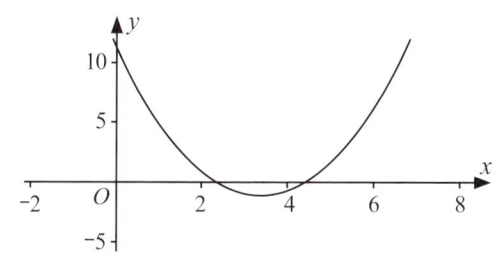

   The gradient is 3 at one of the points:

   (0,11)   (2,1)   (3.5,−1.25)   (5,1)

   At which point is the gradient 3? Explain how you know.

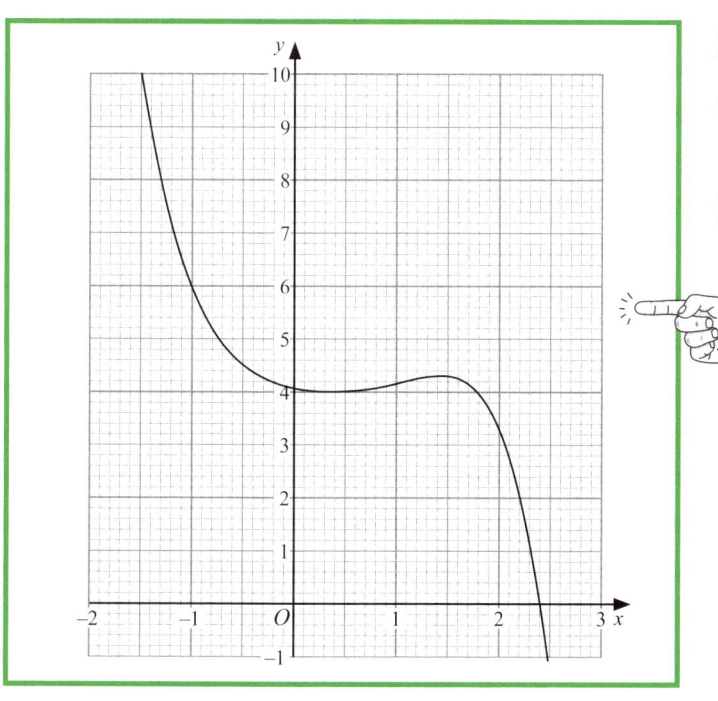

3. The graph of $y = f(x)$ is shown on the grid.

   a) Estimate the gradient of the curve at the point (−1,6).

   b) Find the coordinates of another point on the curve which has the same gradient as the point (−1,6).

READY?

The gradient of a straight line also tells us the **rate of change**. As 'x' increases by 1, the gradient states the change in 'y'.

Gradient = $\frac{3}{2}$ = 1.5

In this example, if 'x' is increased by 1, 'y' increases by 1.5.

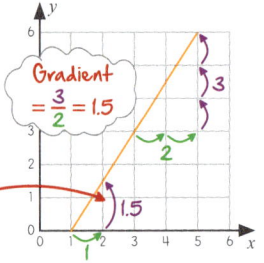

There are many situations where the rate of change provides useful information. For example, the rate of change of distance with respect to time is speed. We can solve rate of change problems involving both straight lines and curves.

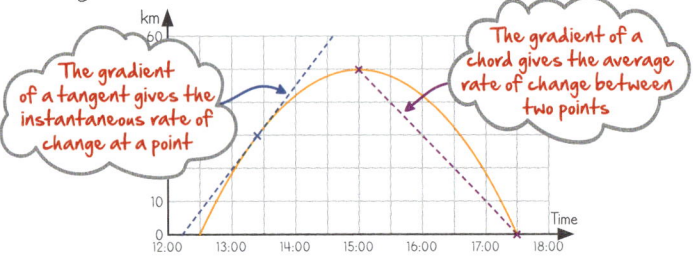

The gradient of a tangent gives the instantaneous rate of change at a point

The gradient of a chord gives the average rate of change between two points

CHECK-IN

1 Estimate the gradient of the tangent when:
$x = 0.4$ $x = 2$

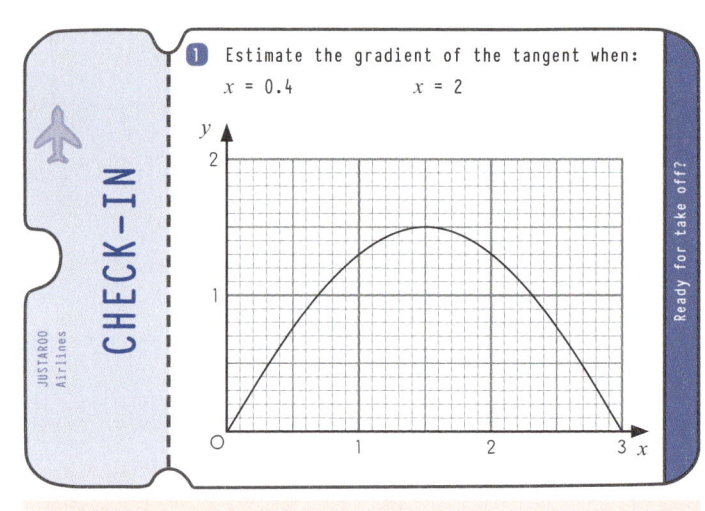

e.g. 1
The graph shows information about the temperature of a cup of coffee in the 8 minutes after it is served. Estimate the rate of change of the temperature at 2 minutes. Describe the meaning of your answer.

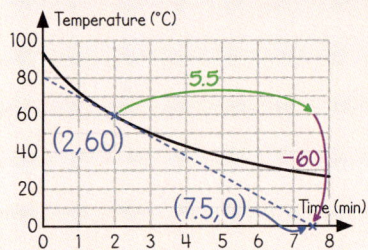

Work out the gradient of the tangent at 2 minutes:

$\frac{\text{change in } y}{\text{change in } x} = \frac{-60}{5.5} = -10.9...$

At two minutes, the temperature is decreasing at a rate of about 10.9°C per minute

e.g. 2
Here is a distance-time graph.

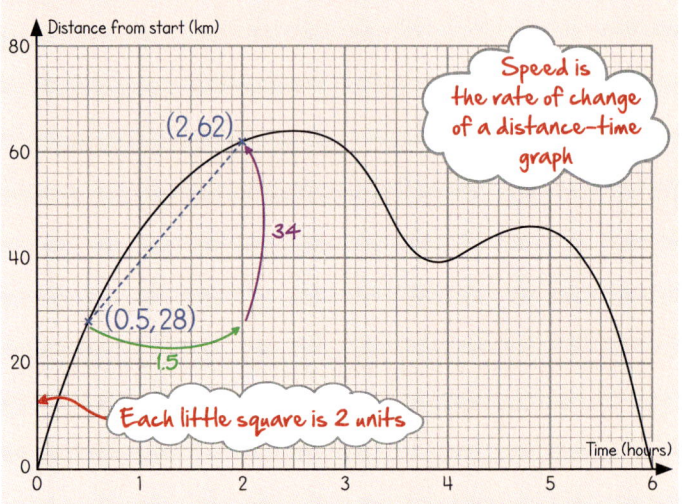

Speed is the rate of change of a distance–time graph

Each little square is 2 units

Estimate the average speed between 30 minutes and two hours. State the units of your answer.

Work out the gradient of the chord between the points at 30 minutes and 2 hours:

$\frac{\text{change in } y}{\text{change in } x} = \frac{34}{1.5} = 22.666...$

The average speed between these times is 22.7 km/h to 1 d.p.

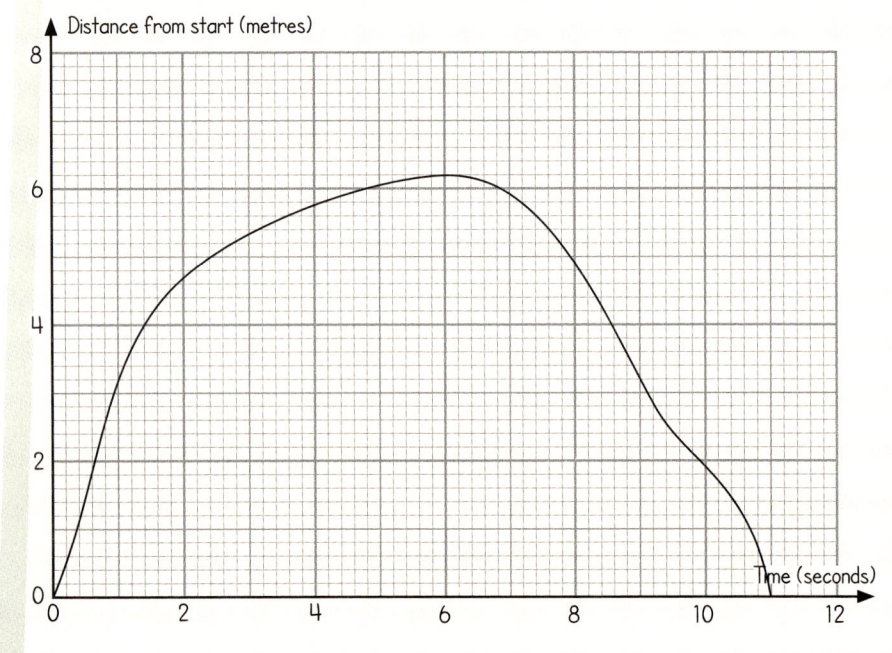

SET?

A. Here is a distance-time graph

(i) Estimate the instantaneous speed at two seconds.

(ii) Estimate the average speed between 6 and 9 seconds.

(iii) Estimate the instantaneous speed at ten seconds.

RATES OF CHANGE

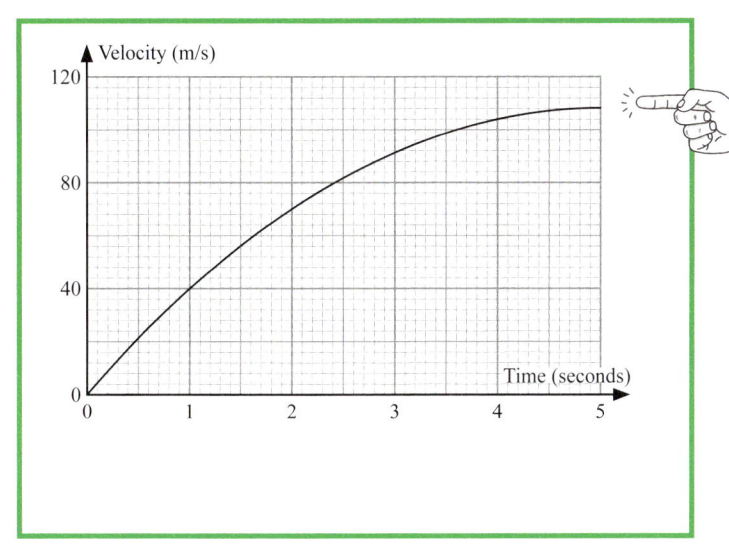

splendid!

1. The velocity-time graph shows information about a skydiver's speed after jumping from an aeroplane.

a) Estimate the rate of change of velocity at one second.

b) Interpret the value you have calculated in part (a).

2. The graph shows the value of an investment throughout a year.

In which month was the rate of change of the value of the investment greatest? Explain your answer.

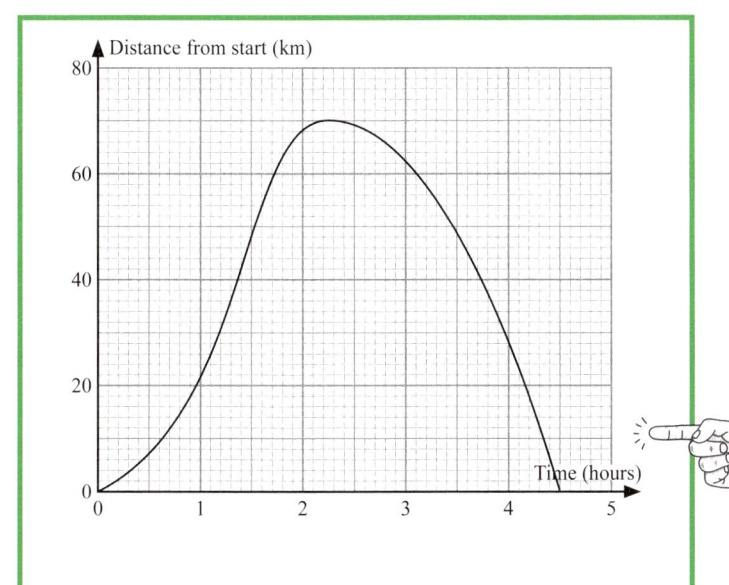

3. Here is a distance-time graph.

a) Estimate the average speed between 30 minutes and 90 minutes. State the units of your answer.

b) Work out the average speed for the whole journey.

c) At what time was the speed the greatest?

d) At what times was the speed 18 kilometres per hour?

READY?

Here is a speed-time graph for a journey:

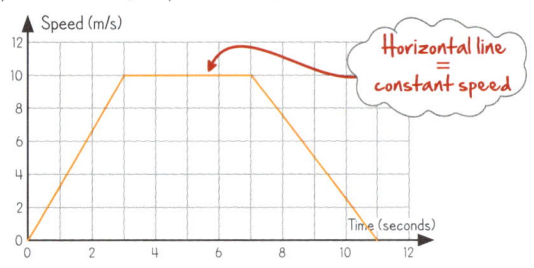

Horizontal line = constant speed

The speed is constant between 3 and 7 seconds. For this part of the journey we can work out the distance travelled using distance = speed × time:

Distance = $10 \times (7 - 3) = 10 \times 4 = 40$ metres

This is the same as the area of the rectangle under the graph between these points: Area = $10 \times 4 = 40$

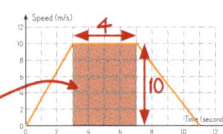

This is true for any shape: the area under a speed-time graph is always equal to distance travelled in that time.

e.g. 1 Work out the distance travelled in the first five hours on this speed-time graph:

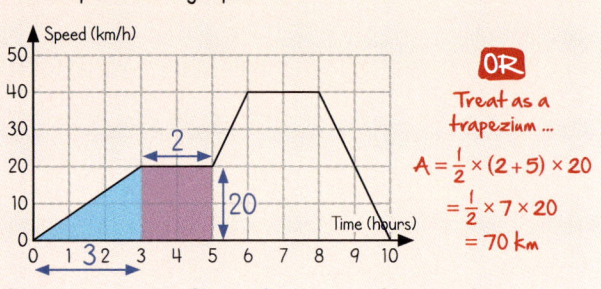

OR

Treat as a trapezium ...
$A = \frac{1}{2} \times (2+5) \times 20$
$= \frac{1}{2} \times 7 \times 20$
$= 70$ km

Distance = area of triangle + area of rectangle
$= \frac{3 \times 20}{2} + 2 \times 20 = 30 + 40 = 70$ km

CHECK-IN

1 Calculate the area of the shapes:

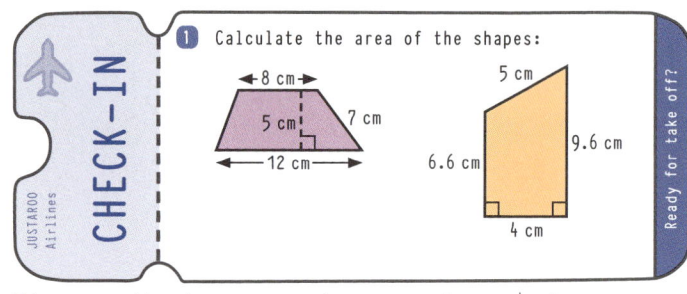

We can estimate areas under curved graphs too ...

e.g. 2 The graph of $y = 5 - x - x^2$ is shown on the grid. Estimate the area under the curve between the points where $x = -2$ and $x = 0$. Use 4 strips of equal width.

1 Draw each strip (often a trapezium)

Each strip is 0.5 wide

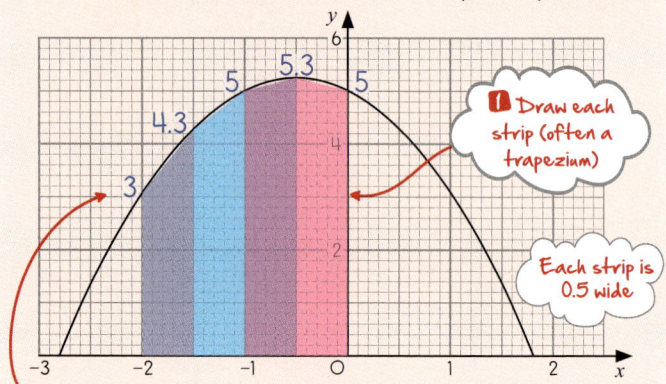

2 Read the y-values off the axis as they are needed for 'a' and 'b'

3 Find the area of each trapezium using $A = \frac{1}{2} \times (a+b) \times h$

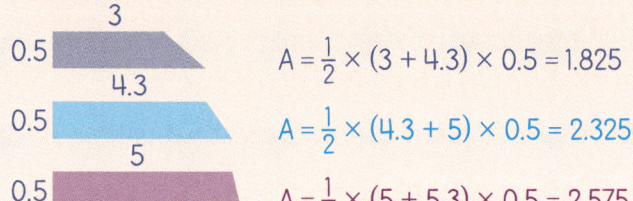

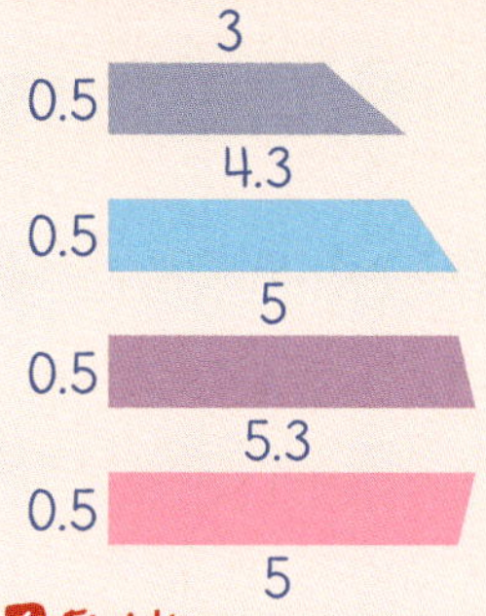

$A = \frac{1}{2} \times (3 + 4.3) \times 0.5 = 1.825$

$A = \frac{1}{2} \times (4.3 + 5) \times 0.5 = 2.325$

$A = \frac{1}{2} \times (5 + 5.3) \times 0.5 = 2.575$

$A = \frac{1}{2} \times (5.3 + 5) \times 0.5 = 2.575$

4 Find the total area: Area = $1.825 + 2.325 + 2.575 + 2.575 = 9.3$

This is an underestimate as each trapezium is below the curve

---

A. Here is a speed-time graph:

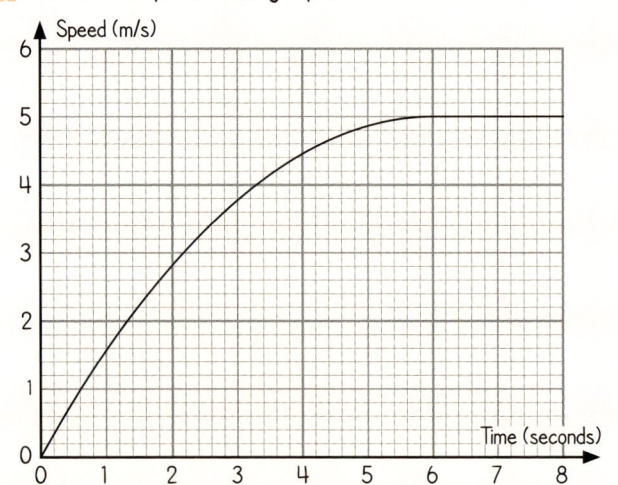

(i) Estimate the distance travelled in the eight seconds.
(ii) Is your answer an overestimate or an underestimate? Give a reason for your answer.

Use 4 strips of equal width

SET?

B. The speed-time graph shows information about an object during a scientific experiment.

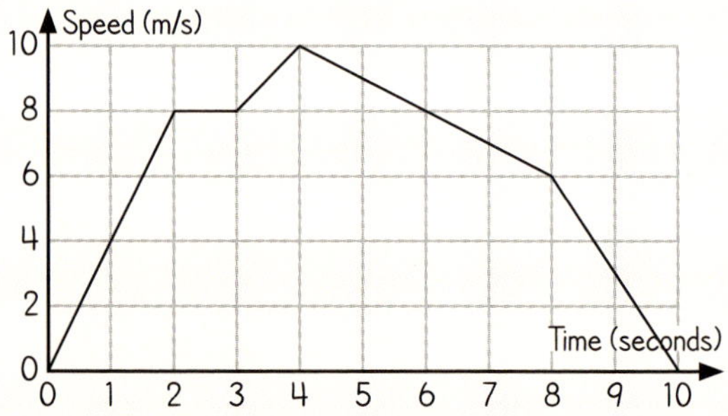

Work out the distance travelled by the object in the first six seconds.

AREA UNDER A GRAPH

1. The graph shows information about the flow of water through a gate on a canal lock.

Bez uses six strips to estimate the area under the graph correctly for the first 40 seconds. He uses his result to write:

"The average flow is 10 575 litres of water per second"

Do you agree with Bez? Explain why.

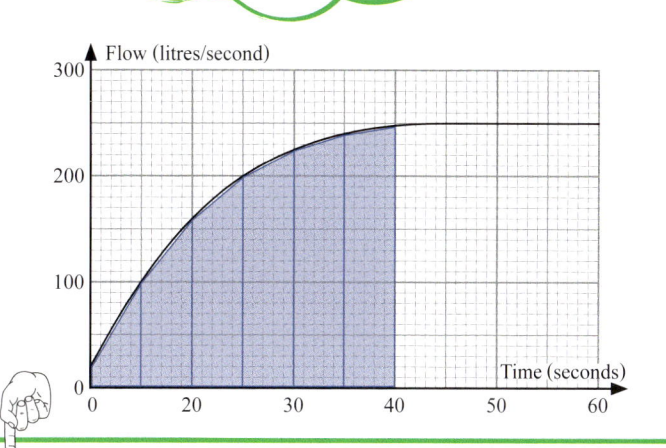

2. The velocity-time graph shows information about a rocket firework.

a) At what time was the acceleration zero?

b) Estimate the acceleration after 2 seconds.

c) Estimate the total distance travelled by the firework. Use strips with a width of 1.

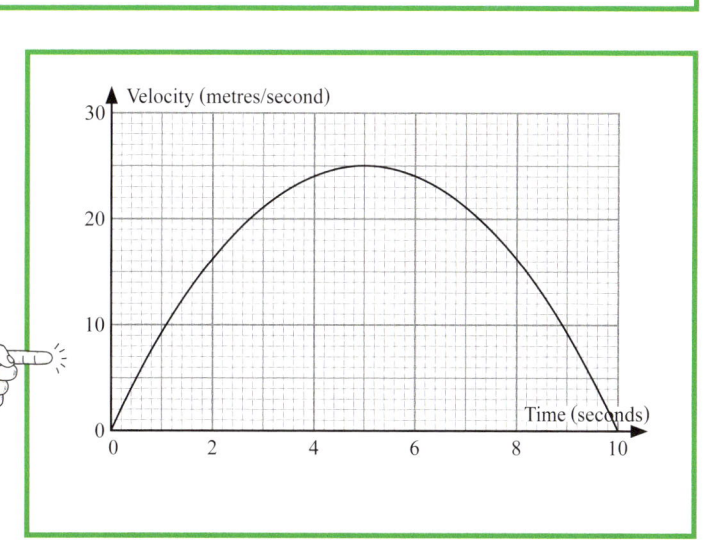

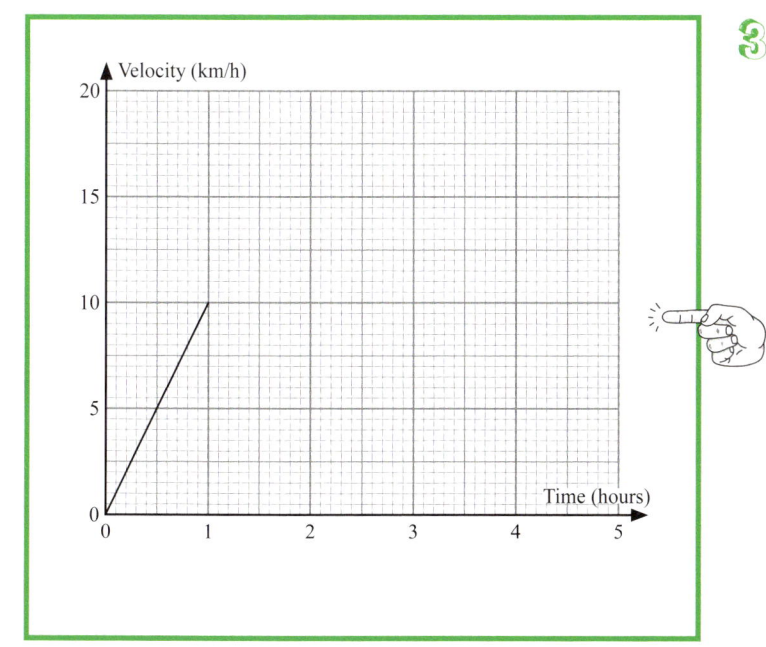

3. The velocity-time graph shows information about the journey of a ship.

The ships accelerates constantly for the first hour. It then travels at a constant velocity for 5 kilometres.

The ship then accelerates again at the same rate to a velocity of 15 km/h.

After travelling at this velocity for one hour, the ship decelerates constantly until stopping.

The total length of the journey is 42.5 kilometres.

Complete the velocity-time graph.

READY?

An equation such as $2x + 9y = 24$ has an infinite number of solutions. If we have a pair of equations that look like this (e.g. $2x + 9y = 24$ and $6x + 2y = 22$) we can find the single solution that fits both. This is known as 'solving simultaneous equations' as we are looking to find when both equations can be solved at the same time.

Simultaneous equations are often solved using a method of elimination. A graph can also be used to identify the solution:

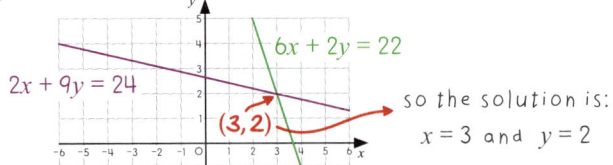

so the solution is: $x = 3$ and $y = 2$

Sometimes it is easier to solve simultaneous equations using the method of substitution ...

e.g. 1 Solve $9x - 4y = 10$ and $y = 6x - 5$

1 Label the equations and check the structure of the equations:
$9x - 4y = 10$ ①
$y = 6x - 5$ ② — This equation is in the form $y = ...$

2 Carry out a substitution to eliminate a variable:
Substitute ② into ①: $9x - 4(6x - 5) = 10$

3 Simplify and solve:
$9x - 24x + 20 = 10$
$-15x + 20 = 10$
$-15x = -10$
$x = \frac{-10}{-15} = \frac{10}{15} = \frac{2}{3}$

4 Substitute $x = \frac{2}{3}$ into ② (or ①) to find the value of y:
$y = 6 \times \frac{2}{3} - 5$
$= 4 - 5$
$= -1$

5 Write the solution: $x = \frac{2}{3}$ and $y = -1$

CHECK-IN

Solve the simultaneous equations:

1 $2x + 7y = 38$
    $5x + 3y = 37$

2 $3x + 2y = 4$
    $4x - 5y = -33$

3 $5a - 8b = 19$
    $6a + 4b = -18$

4 $2p - 8q = 2$
    $5p - 6q = 12$

Simultaneous equations may involve fractions. Lots of possible routes can lead to the same solution ...

e.g. 2 Solve $\frac{2}{3}x - \frac{4}{5}y = 4$ and $\frac{1}{2}x + \frac{2}{3}y = -16$

Eliminate any fractions by multiplying each equation:

$\times 15 \left(\frac{2}{3}x - \frac{4}{5}y = 4 \right) \times 15$ → $10x - 12y = 60$

$\times 6 \left(\frac{1}{2}x + \frac{2}{3}y = -16 \right) \times 6$ → $3x + 4y = -96$

LCM of the denominators

Label the equations, then check coefficients:
$10x - 12y = 60$ ①
$3x + 4y = -96$ ②

② × 3 → $9x + 12y = -288$ ③
$10x - 12y = 60$ ①

Both 'y's now have a coefficient of 12, but with a different operation (+ and −) ... so add the equations to find x

③ + ① → $19x = -228$
÷19   ÷19
$x = -12$

Substitute $x = -12$ into ① (or ②) to find the value of y:
$10 \times -12 - 12y = 60$
$-120 - 12y = 60$
$+120$ $+120$
$-12y = 180$
÷−12 ÷−12
$y = -15$

Check:
①: $10 \times -12 - 12 \times -15$
$= 60$ ✓
and
②: $3 \times -12 + 4 \times -15$
$= -96$ ✓

The solution is: $x = -12$ and $y = -15$

SET?

Solve the simultaneous equations:

A. $3x + 7y = 85$
   $y = 5x - 53$

B. $10x - 3y = 64$
   $y = 3x - 20$

C. $2p + 15q = -6$
   $p = 5q - 8$

D. $\frac{1}{2}x - \frac{3}{4}y = 10$
   $\frac{3}{4}x - \frac{3}{8}y = 3$

E. $\frac{3}{4}a + \frac{1}{2}b = -3$
   $\frac{4}{5}a - 3b = -35$

F. $\frac{3}{5}x - \frac{1}{4}y = \frac{9}{20}$
   $9x + 25y = 1$

GO! SIMULTANEOUS EQUATIONS 1

1. The graphs of $\frac{5}{16}x + \frac{7}{3}y = 3$ and $y = \frac{7}{8}x - 2$ are shown on the grid.

Estimate the solution of the simultaneous equations $\frac{5}{16}x + \frac{7}{3}y = 3$ and $y = \frac{7}{8}x - 2$

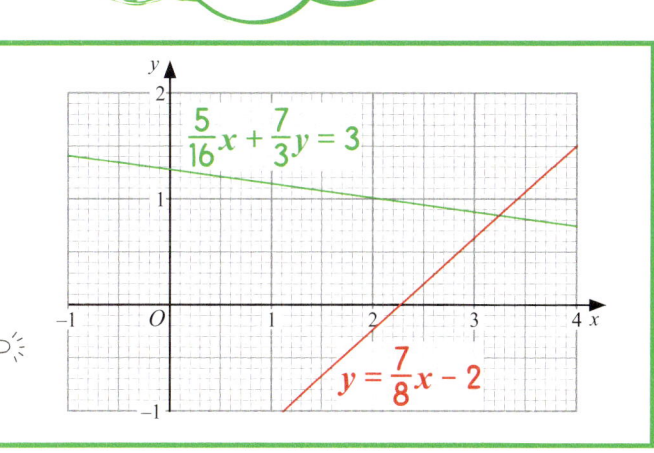

2. Nargis is solving the simultaneous equations $\frac{2}{5}a - 3b = -28$ and $3a + 5b = 120$

She starts by writing:

$\frac{2}{5}a - 3b = -28$
×5 ↓ ↓ ×5
$10a - 15b = -140$

Explain the mistake that Nargis has made.

3. Loz is solving the simultaneous equations $5x - 2y = -3$ and $y = 3x - 2$. He writes:

$5x - 2(3x - 2) = -3$
$5x - 6x - 4 = -3$
$-x - 4 = -3$
$-x = 1$
$x = -1$ so $y = -5$

Explain the mistake Loz has made.

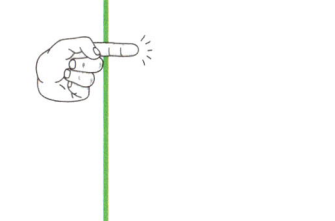

4. Solve the simultaneous equations:

$\frac{4a}{3} + \frac{3b}{5} = -\frac{19}{5}$

$\frac{3a}{4} + \frac{b}{2} = -\frac{25}{12}$

READY?

If we plot the graphs of $y = 2x - 2$ and $y = 2x^2 - 5x + 1$ on the same grid, we can see that they intersect at the points $(3, 4)$ and $(0.5, -1)$.

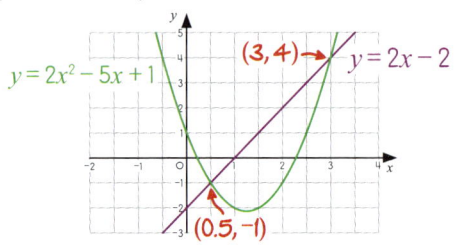

This means that the simultaneous equations $y = 2x - 2$ and $y = 2x^2 - 5x + 1$ have two solutions:
$$x = 3, y = 4 \text{ and } x = 0.5, y = -1$$

We also need to solve pairs of equations algebraically when one is linear and one is quadratic.

CHECK-IN

1. Solve $5x - 4y = 12$ and $y = 3x - 10$
2. Solve $2x^2 + 7x - 9 = 0$
3. Use the formula for solving quadratic equations to solve: $6x^2 + 4x - 3 = 0$. Give the solutions to two decimal places.

Ready for take off?

---

e.g. 1 Solve $y = x^2 - 3x - 10$ and $y = 3x - 3$

1. **Label the equations:**
   $y = x^2 - 3x - 10$ ①
   $y = 3x - 3$ ②

2. **Carry out a substitution to eliminate a variable:**
   Substitute ① into ②: $x^2 - 3x - 10 = 3x - 3$

3. **Rearrange into the form $ax^2 + bx + c = 0$:**
   $x^2 - 3x - 10 = 3x - 3$
   $x^2 - 6x - 10 = -3$
   $x^2 - 6x - 7 = 0$

   This quadratic factorises

4. **Solve:**
   $(x + 1)(x - 7) = 0$
   $x = -1$ and $x = 7$

5. **Substitute each value of x into ② (or ①) to find the corresponding value of y:**
   $x = -1$: $y = 3 \times -1 - 3 = -6$
   $x = 7$: $y = 3 \times 7 - 3 = 18$

6. **Write the solution:**
   $x = -1, y = -6$ and $x = 7, y = 18$

---

e.g. 2 Solve $3b - a^2 = 12$ and $a + b = 10$

1. **Rearrange if needed, and label the equations:**
   $3b - a^2 = 12$ ① and $a + b = 10 \to b = 10 - a$ ②

2. **Substitute** ② into ①: $3(10 - a) - a^2 = 12$

3. **Rearrange:**
   $30 - 3a - a^2 = 12$
   $18 - 3a - a^2 = 0$ ×-1
   $a^2 + 3a - 18 = 0$

   Can factorise

4. **Solve:**
   $(a + 6)(a - 3) = 0$
   $a = -6$ and $a = 3$

5. **Substitute into** ②:
   $a = -6$: $b = 10 - -6 = 16$
   $a = 3$: $b = 10 - 3 = 7$

6. **Write the solution:**
   $a = -6, b = 16$ and $a = 3, b = 7$

---

e.g. 3 Solve $x^2 + 2y^2 = 11$ and $2x + y = 1$. Give the solutions to two decimal places.

1. $x^2 + 2y^2 = 11$ ① and $2x + y = 1 \to y = 1 - 2x$ ②

2. **Substitute** ② into ①: $x^2 + 2(1 - 2x)^2 = 11$

3. **Rearrange:**
   $x^2 + 2(1 - 4x + 4x^2) = 11$
   $x^2 + 2 - 8x + 8x^2 = 11$
   $9x^2 - 8x - 9 = 0$

4. **Solve, using** $x = \dfrac{-b \pm \sqrt{b^2 - 4ac}}{2a}$
   $x = \dfrac{8 \pm \sqrt{(-8)^2 - 4 \times 9 \times -9}}{2 \times 9}$
   $x = 1.538...$ and $x = -0.649...$

5. **Substitute into** ②:
   $x = 1.538... \to y = -2.077...$
   and
   $x = -0.649... \to y = 2.299...$

6. **Write the solution**
   $x = 1.54, y = -2.08$ and
   $x = -0.65, y = 2.30$
   to 2 decimal places

Solve the simultaneous equations:

A. $y = x^2 - 5x - 3$
$y = 9 - x$

B. $y = 3x^2 - 10x - 8$
$-2x + y = 7$

Solve, giving solutions to two decimal places:

C. $y = 4x^2 + 5x - 1$
$y = 4 - 2x$

D. $5x - 4y^2 = 7$
$y = 2x - 9$

E. $4x^2 + y^2 = 13$
$y = x + 3$

SET?

68

GO! SIMULTANEOUS EQUATIONS 2

Wonderful!

1. Dale is asked to solve the simultaneous equations:
   $$y = x^2 + 8x - 5$$
   $$2x + y = 6$$

   He writes the answer:
   $$x = 1 \text{ and } y = 4$$

   Do you agree with Dale? Explain why.

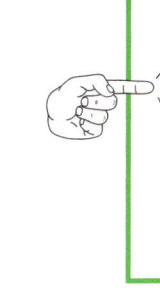

2. The graphs of $y = -2x^2 + 4x + 5$ and $y = 8 - 3x$ are shown on the grid. 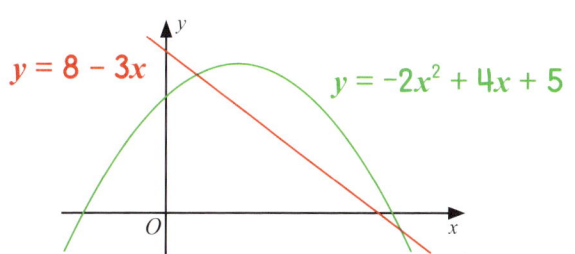

   Find the coordinates of the points of intersection of the two graphs.

3. The area of the rectangle is equal to the area of the triangle. 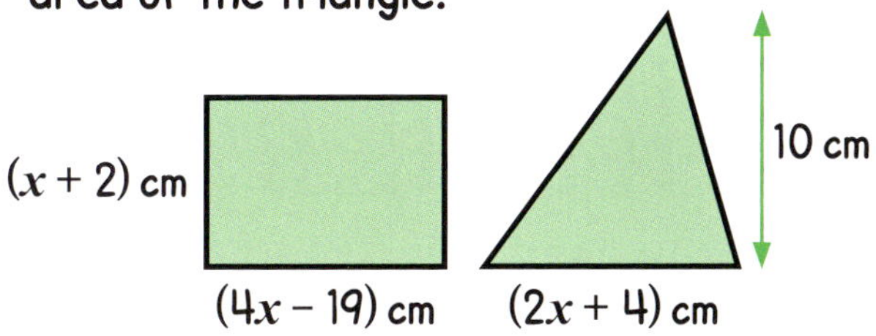

   Rectangle: $(x + 2)$ cm by $(4x - 19)$ cm
   Triangle: base $(2x + 4)$ cm, height 10 cm

   Find the perimeter of the rectangle.

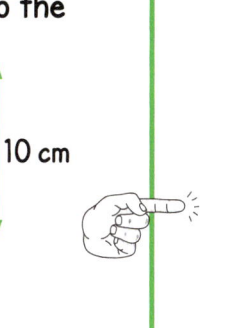

4. Find the exact solutions of the simultaneous equations:
   $$3x^2 + y^2 = 57$$
   $$2y = x + 2$$

READY?

When we are changing the subject of a formula it sometimes helps to use a function machine.
For example, if $y = 4x + 3$ we can see that $x = \frac{y-3}{4}$

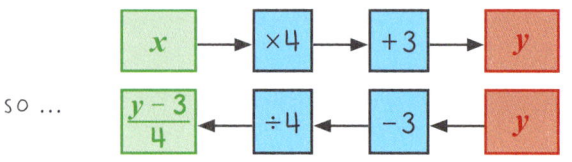

$x \to \times 4 \to +3 \to y$

so... $\frac{y-3}{4} \to \div 4 \to -3 \to y$

However, if the required subject appears twice, we need to approach the problem differently...

e.g. 1 Make e the subject of $A = \frac{2e-3}{e}$

1. It is usually helpful to eliminate fractions from the formula:
$A = \frac{2e-3}{e}$
$\times e \quad \times e$ — *Multiply both sides by e*

2. Get all terms involving e onto the same side:
$Ae = 2e - 3$
$-2e \quad -2e$

3. Remove the common factor of e:
$Ae - 2e = -3$
$e(A - 2) = -3$ — *Factorise*
$\div (A-2) \quad \div (A-2)$

4. Isolate e: $e = \frac{-3}{A-2}$

This could be written differently, e.g.: $e = -\frac{3}{A-2}$ **OR** $e = \frac{3}{2-A}$

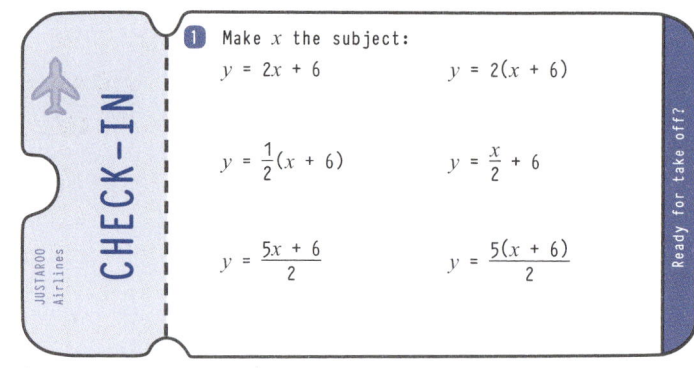

CHECK-IN

① Make x the subject:

$y = 2x + 6$ $\qquad y = 2(x+6)$

$y = \frac{1}{2}(x+6)$ $\qquad y = \frac{x}{2} + 6$

$y = \frac{5x+6}{2}$ $\qquad y = \frac{5(x+6)}{2}$

Sometimes more than one step is needed in order to get all terms with the variable on one side...

e.g. 2 Rearrange $a = \frac{5b+1}{4-b}$ to make b the subject.

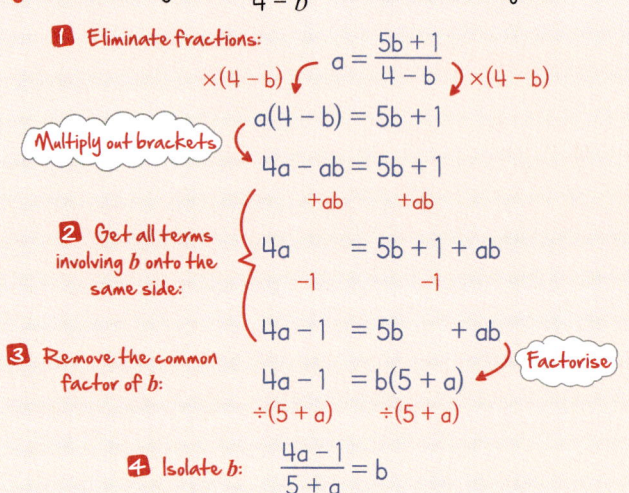

1. Eliminate fractions:
$\times (4-b) \quad a = \frac{5b+1}{4-b} \quad \times (4-b)$
$a(4-b) = 5b + 1$
Multiply out brackets
$4a - ab = 5b + 1$
$+ab \qquad +ab$

2. Get all terms involving b onto the same side:
$4a = 5b + 1 + ab$
$-1 \qquad -1$
$4a - 1 = 5b + ab$

3. Remove the common factor of b:
$4a - 1 = b(5 + a)$ — *Factorise*
$\div (5+a) \quad \div (5+a)$

4. Isolate b: $\frac{4a-1}{5+a} = b$

This could be written differently, e.g.: $b = \frac{4a-1}{a+5}$

Not all rearranging problems ask us to 'change the subject', but the balancing approach is always helpful:

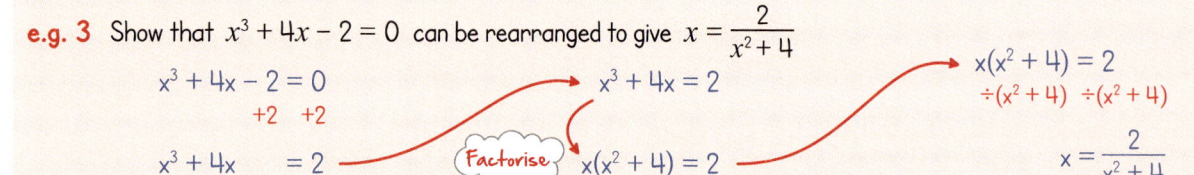

e.g. 3 Show that $x^3 + 4x - 2 = 0$ can be rearranged to give $x = \frac{2}{x^2+4}$

$x^3 + 4x - 2 = 0$
$+2 \quad +2$
$x^3 + 4x = 2 \quad \to \quad$ *Factorise* $\quad x(x^2+4) = 2 \quad \to \quad x(x^2+4) = 2$
$\div (x^2+4) \quad \div(x^2+4)$
$x = \frac{2}{x^2+4}$

SET?

A. Make x the subject:

(i) $y = \frac{3x+1}{x}$

(ii) $y = \frac{4x-3}{x}$

(iii) $y = \frac{ax+b}{x}$

(iv) $y = \frac{7-4x}{x}$

(v) $y = \frac{6x+1}{9x}$

B. Make a the subject:

(i) $T = \frac{a+12}{a-7}$

(ii) $b = \frac{a+20}{6+a}$

(iii) $d = \frac{ax-bx}{a+b}$

(iv) $A = \left(\frac{a+b}{2}\right)h$

C. Show that $x^3 - 6x + 23 = 0$ can be written in the form $x = \frac{x^3+23}{6}$

D. Show that $x^3 - 4x^2 - 7 = 0$ can be written in the form $x = \frac{7}{x^2} + 4$

GO! REARRANGING FORMULAE

Exceptional!

1. Show that $x^3 - 2x - 9 = 0$ can be rearranged to give:

a) $x = \sqrt[3]{2x + 9}$

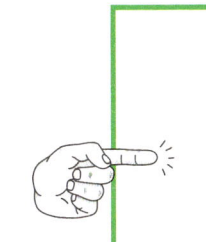

b) $x = \dfrac{x^3 - 9}{2}$

c) $x = \dfrac{9}{x^2 - 2}$

2. Show that $x^2 - 3 = \dfrac{4}{x}$ can be rearranged to give $x = \sqrt[3]{3x + 4}$

3. Make a the subject of $s = ut + \dfrac{1}{2}at^2$

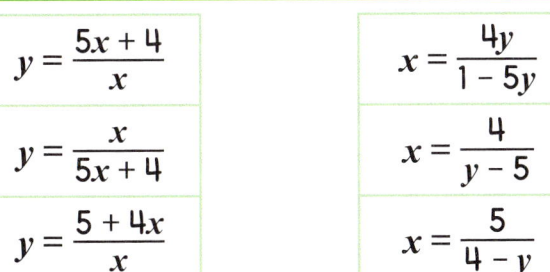

4. By rearranging, match the formulae that are equivalent.

$y = \dfrac{5x + 4}{x}$	$x = \dfrac{4y}{1 - 5y}$
$y = \dfrac{x}{5x + 4}$	$x = \dfrac{4}{y - 5}$
$y = \dfrac{5 + 4x}{x}$	$x = \dfrac{5}{4 - y}$
$y = 4 - \dfrac{5}{x}$	$x = \dfrac{5}{y - 4}$

5. Make p the subject of $\dfrac{1}{p} + \dfrac{2}{q} = \dfrac{3}{r}$

READY?

Some equations cannot be solved by standard algebraic methods. It is often possible to estimate solutions to these equations instead. E.g. the solution to $x^3 + 4x - 8 = 0$ can be visualised using the graph of $y = x^3 + 4x - 8$

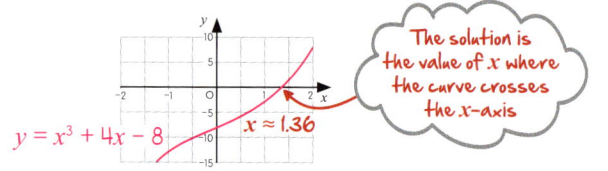

$y = x^3 + 4x - 8$

The solution is the value of x where the curve crosses the x-axis

$x \approx 1.36$

We can also substitute values and compare outputs:

e.g. 1 Show that the equation $x^3 + 4x - 8 = 0$ has a solution between $x = 1$ and $x = 2$.

When $x = 1$: $1^3 + 4 \times 1 - 8 = 1 + 4 - 8 = -3$

When $x = 2$: $2^3 + 4 \times 2 - 8 = 8 + 8 - 8 = 8$

Always write a conclusion ...

Since 0 is between -3 and 8, x must be between 1 and 2

Sometimes it helps to rearrange the equation first:

e.g. 2 Show that the equation $x^3 + 3x^2 = 1$ has a solution between $x = -2.9$ and $x = -2.8$.

Sign-change method:

❶ Rearrange the equation to make 0 on one side:
$x^3 + 3x^2 = 1 \rightarrow x^3 + 3x^2 - 1 = 0$

❷ Use the updated equation to continue as in e.g. 1:

When $x = -2.9$: $(-2.9)^3 + 3 \times (-2.9)^2 - 1 = -0.159$

When $x = -2.8$: $(-2.8)^3 + 3 \times (-2.8)^2 - 1 = 0.568$

Since 0 is between -0.159 and 0.568 (a change of sign) x must be between -2.9 and -2.8

CHECK-IN

❶ Evaluate when $x = -3.2$:
$x^3 - 5x + 3$ $x^3 + 3x^2 - 6$

❷ Estimate the solutions of $\frac{1}{2}x^2 + 3x + 1 = 0$:

$y = \frac{1}{2}x^2 + 3x + 1$

Any process of repetition towards a solution is called an *iterative process*. Another way to estimate solutions is by using an iterative formula:

e.g. 3 The equation $x^3 - 5x + 3 = 0$ has a solution between $x = 0$ and $x = 1$. Starting with $x_0 = 1$, use the iterative formula:

$$x_{n+1} = \frac{x_n^3 + 3}{5}$$

four times to estimate the value of this solution correct to two decimal places.

An iterative formula allows us to take an existing estimate and work out the next best estimate. We have been given a starting value.

❶ Enter the starting value into the calculator: 1 = or EXE
$x_0 = 1$

❷ Enter the iterative formula into the calculator using Ans:

[÷] [Ans] [☐²] 3 [>] + 3 [∨] 5

❸ Press = (or EXE) repeatedly, and record the output each time (we may need S⇔D or FORMAT too):

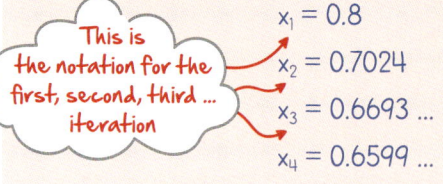

$x_1 = 0.8$
$x_2 = 0.7024$
$x_3 = 0.6693 ...$
$x_4 = 0.6599 ...$

This is the notation for the first, second, third ... iteration

The values are 'homing in' on the solution

❹ Write the solution as required:
$x = 0.66$ to 2 decimal places

A. Show that a solution to:
$x^3 - 6x + 23 = 0$
lies between $x = -4$ and $x = -3$

B. Show that a solution to:
$x^3 + 5x - 13 = 0$
lies between $x = 1.6$ and $x = 1.7$

C. Show that a solution to:
$x^3 - 4x^2 = 7$
lies between $x = 4$ and $x = 5$

D. The equation $x^3 - 6x - 4 = 0$ has a solution between $x = -1$ and $x = 0$. Use the iterative formula:

$$x_{n+1} = \frac{x_n^3}{6} - \frac{2}{3}$$

five times to estimate the solution to the equation. Start with $x_0 = 2$ and give your answer to two decimal places.

SET?

E. The iterative formula:

$$x_{n+1} = \sqrt[3]{5x_n^2 - 8}$$

can be used to find an estimate for the solution to the equation $x^3 - 5x^2 + 8 = 0$

Starting with $x_0 = 4$, work out the value of x_3 to three decimal places.

GO! ITERATION

1. a) Show that the equation $x^2 - 2x^3 + 4 = 0$ has a solution between $x = 1$ and $x = 2$

b) Show that $x^2 - 2x^3 + 4 = 0$ can be rearranged to give: $x = \sqrt{\dfrac{4}{2x-1}}$

c) Using $x_0 = 1.5$ and the iterative formula:
$$x_{n+1} = \sqrt{\dfrac{4}{2x_n - 1}}$$
four times, find an estimate for a solution to $x^2 - 2x^3 + 4 = 0$

2. The equation $4x^3 - 2x^2 = 3$ has one solution. Find two consecutive numbers to one decimal place between which the solution lies.

3. The number of butterflies on a nature reserve is modelled using the iterative formula: $B_{n+1} = 1.035(B_n - 20)$ where B_n is the number of butterflies on the reserve in June of year n.

There were 120 butterflies on the reserve in June of 2021. What is the first year in which the model predicts that there will be no butterflies remaining?

4. Using the iterative formula $x_{n+1} = \dfrac{14}{x_n^2 + 7}$ with $x_0 = 2$:

a) Work out the values $x_1, x_2, x_3,$ and x_4

b) Find the solution to $x^3 + 7x - 14 = 0$ correct to two decimal places. Justify your answer.

READY?

The inequality $x > 1$ can shown on a number line:

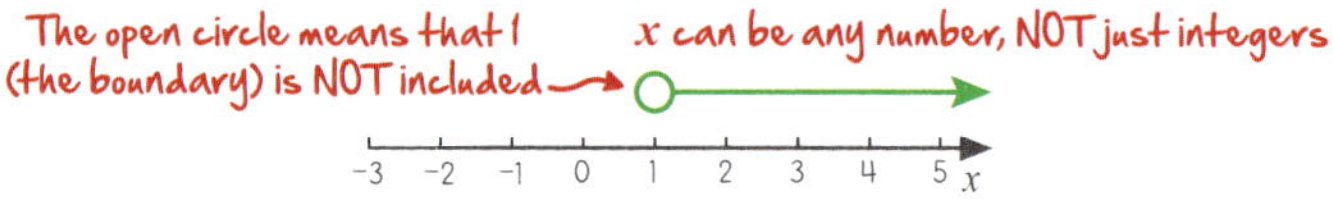

The open circle means that 1 (the boundary) is NOT included → ○
x can be any number, NOT just integers

We can also show $x > 1$ by shading a region on a graph. The dotted line $x = 1$ shows that the boundary is NOT included.

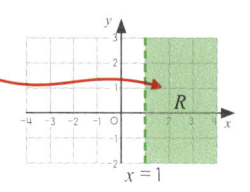

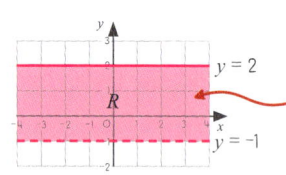

The inequality $-1 < y \leq 2$ can also be represented by a region on a graph. The solid line $y = 2$ shows that the boundary is included.

A graph can show inequalities with two variables:

e.g. 1 Shade the region that satisfies $y \geq 2x + 1$. Label the region R.

1. Draw the (solid) boundary line at $y = 2x + 1$
2. Test a point to check which side of the line is needed; e.g.
   At $(3,4)$ $x = 3$ and $y = 4$
   $2x + 1 \rightarrow 2 \times 3 + 1 = 7$
   Is $y \geq 2x + 1$? $4 \geq 7$ ✗
   At $(0,3)$ $x = 0$ and $y = 3$
   $2x + 1 \rightarrow 2 \times 0 + 1 = 1$
   Is $y \geq 2x + 1$? $3 \geq 1$ ✓ so the required region is above the line
3. Shade and label the region

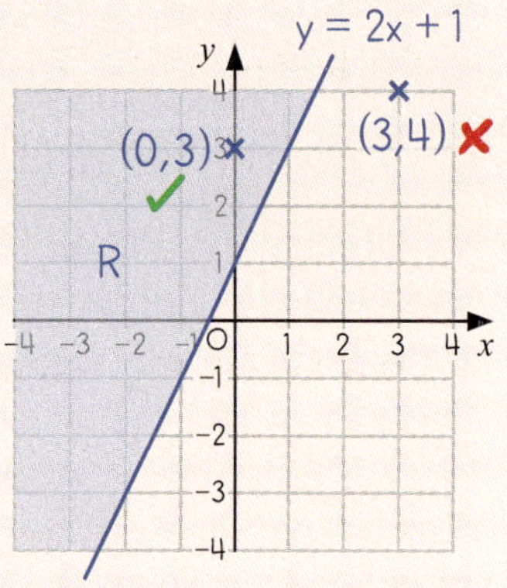

CHECK-IN

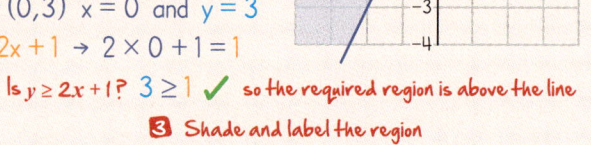

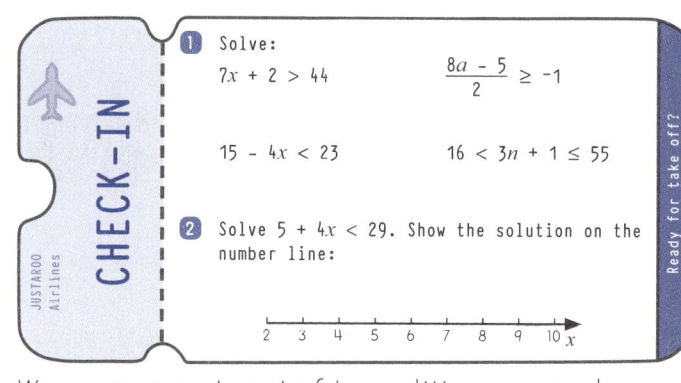

1. Solve:
   $7x + 2 > 44$
   $\dfrac{8a - 5}{2} \geq -1$
   $15 - 4x < 23$
   $16 < 3n + 1 \leq 55$

2. Solve $5 + 4x < 29$. Show the solution on the number line:

We can represent a set of inequalities on a graph...

e.g. 2 Find the region that is defined by the inequalities:
$$y \leq \tfrac{1}{2}x - 1 \qquad y \geq -2 \qquad x + y < 3$$
Label the region R.

Consider each inequality separately. It can help to shade the regions that are NOT needed.

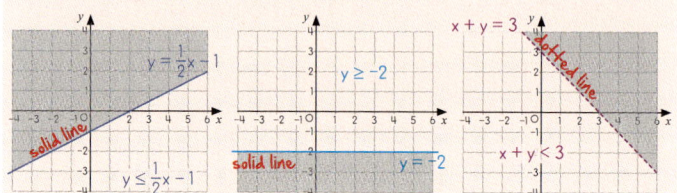

Combine the three inequalities on one graph:

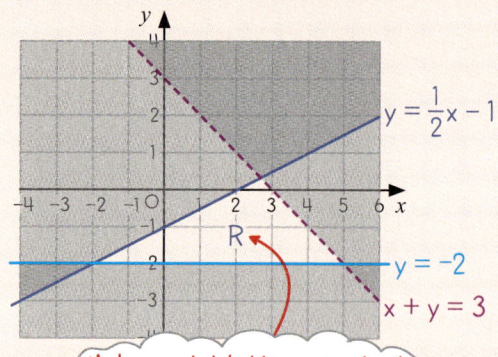

Make sure to label the region clearly

A. Shade the region that satisfies the inequality:
$$y > 3x + 2$$
Label the region R.

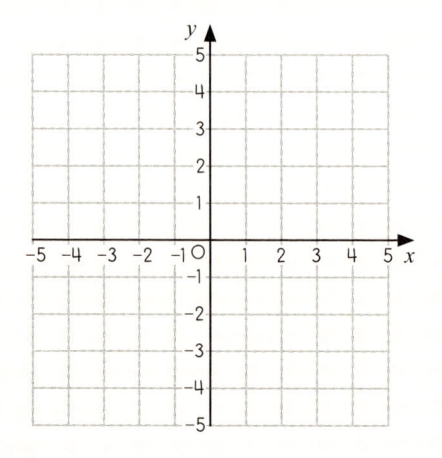

B. Find the region that is defined by the inequalities:
$$x - y \leq 4 \qquad x < 3 \qquad y \leq 3x - 5$$
Label the region R.

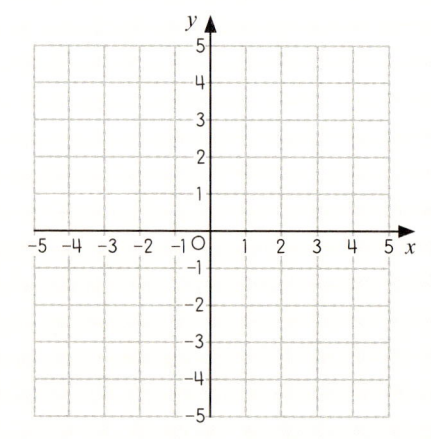

SET?

C. Find the region that is defined by the inequalities:
$$y > -2 \qquad x \geq -3 \qquad x + 2y \leq 1$$
Label the region R.

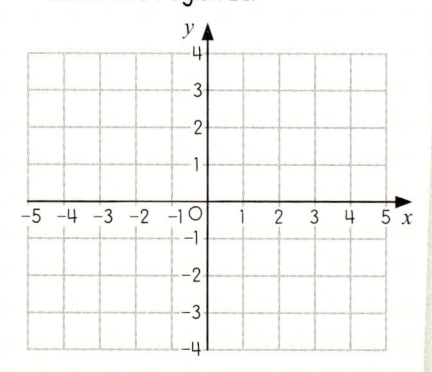

74

GO! INEQUALITIES 1

Grand!

1. The region R satisfies the inequalities:

$y \leq \frac{1}{2}x + 2$

$2x - y < 3$

$x \geq -4$

Use shading to show the region on the grid.

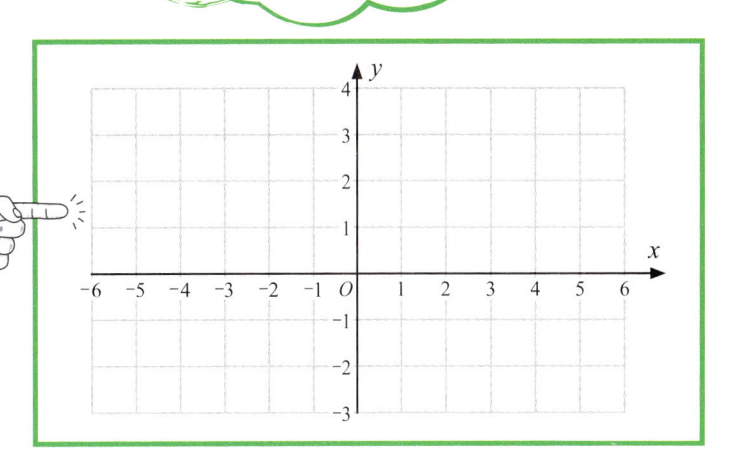

2. Lucus is asked to list the points in the shaded region whose coordinates are integers. He writes:

(−2,2)   (−1,2)   (0,2)   (1,2)

(2,2)   (−1,1)   (0,1)   (−1,0)

Lucus is wrong. Explain why.

3. Use inequalities to describe the region that is labelled R.

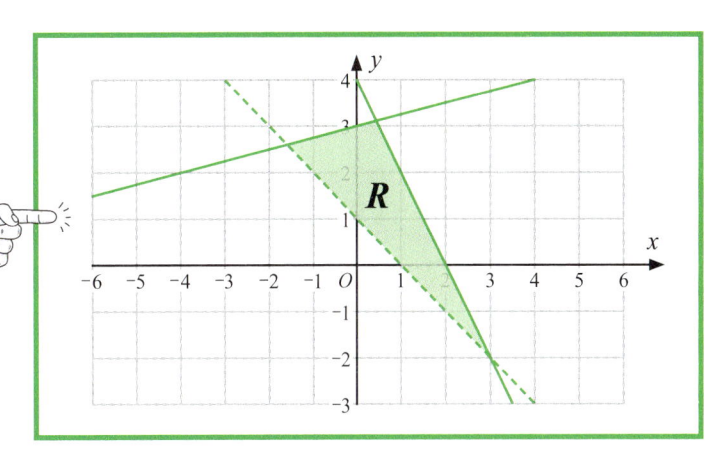

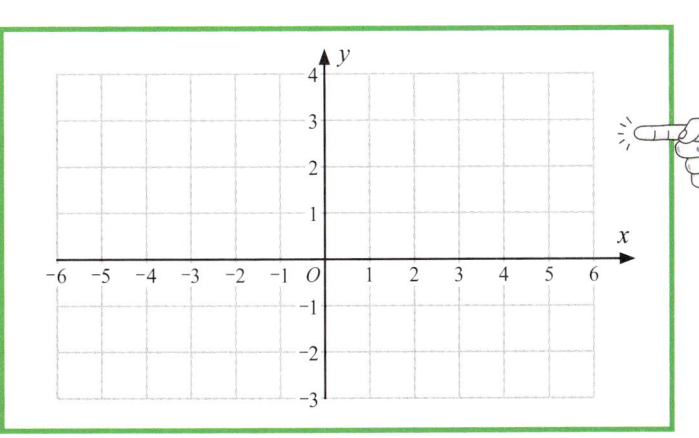

4. A region is defined by the inequalities:

$x > -1$

$y \geq 2x - 1$

$y \leq 3 - \frac{1}{2}x$

List every point in the region whose coordinates are integers.

READY?

If we solve the equation $x^2 = 4$ there are two solutions:
$$x = -2 \text{ and } x = 2$$
This can be shown on a graph:

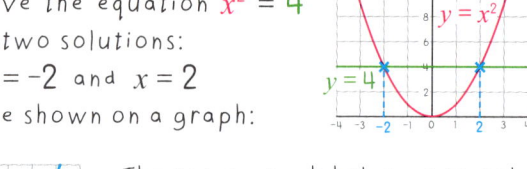

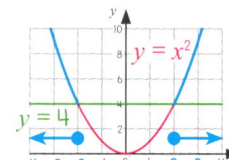

The same graph helps us to solve the inequality $x^2 \geq 4$. We can see that the solution is:
$$x \leq -2 \text{ and } x \geq 2$$

However, if the inequality was $x^2 \leq 4$ the solution would be:
$$-2 \leq x \leq 2$$

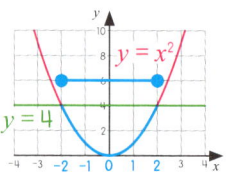

A graph is always helpful when solving quadratic inequalities

e.g. 1 Solve the inequality $x^2 - 2x - 3 < 0$

1 Find the boundary points by considering $x^2 - 2x - 3 = 0$
if $x^2 - 2x - 3 = 0$ then $(x - 3)(x + 1) = 0$
so $x = 3$ and $x = -1$

2 Sketch the graph of $y = x^2 - 2x - 3$

3 Use the graph to visualise when $x^2 - 2x - 3 < 0$
This will be below the x-axis (the line $y = 0$)

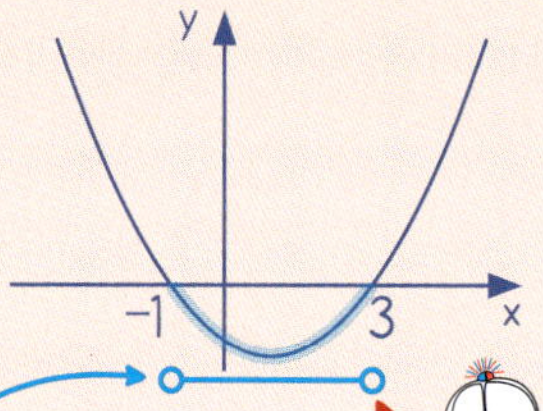

4 Write the solution: so $-1 < x < 3$
The boundary points are NOT included

CHECK-IN

1. Solve $x^2 + 6x - 7 = 0$
2. Solve $2x^2 - 3x - 35 = 0$
3. Sketch the graph of $y = -x^2 - 5x + 36$. Label any points of intersection with the axes.

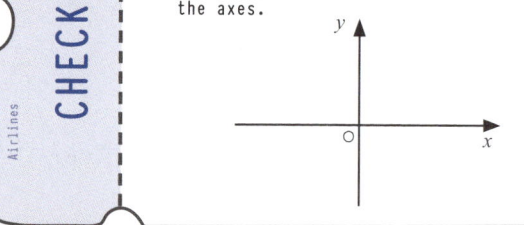

Ready for take off?

e.g. 2 Solve $3x^2 - 17x + 10 \geq 0$

1 Find the boundary points: if $3x^2 - 17x + 10 = 0$
then $(3x - 2)(x - 5) = 0$ so $x = \frac{2}{3}$ and $x = 5$

2 Sketch the graph of $y = 3x^2 - 17x + 10$

3 Use the graph to visualise when $3x^2 - 17x + 10 \geq 0$ 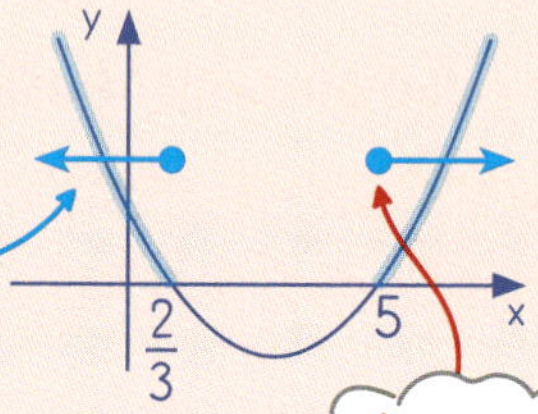
Above the x-axis

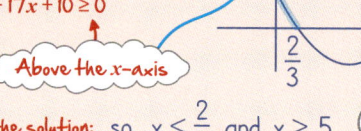

4 Write the solution: so $x \leq \frac{2}{3}$ and $x \geq 5$
Boundary points included

e.g. 3 Solve $20 + x - x^2 < 0$

1 Find the boundary points: if $20 + x - x^2 = 0$
then $(5 - x)(x + 4) = 0$ so $x = 5$ and $x = -4$

2 Sketch the graph of $y = 20 + x - x^2$

3 Use the graph to visualise when $20 + x - x^2 < 0$

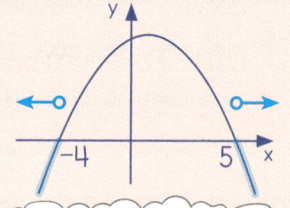

4 Write the solution: so $x < -4$ and $x > 5$
The negative x^2 term means the graph looks like:

SET?

A. Solve $x^2 + x - 42 < 0$

B. Solve $x^2 + 6x + 5 \geq 0$

C. Solve $x^2 - 14x + 48 \leq 0$

D. Solve $2x^2 - 21x + 40 \geq 0$

E. Solve $5x^2 - 7x - 6 > 0$

F. Solve $-x^2 - 6x + 16 \leq 0$

G. Solve $-x^2 + 2x + 8 < 0$

H. Solve $6x^2 + 19x + 15 \leq 0$

76

GO! INEQUALITIES 2

1. Muriel is solving the inequality $4g^2 - 25 < 0$

She writes:
$$4g^2 - 25 < 0$$
$$4g^2 < 25$$
$$g^2 < \frac{25}{4}$$
so $g < \frac{5}{2}$

Explain why Muriel is not correct.

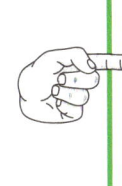

2. Write the set of integers that satisfy the inequality:
$$77 - 17p - 4p^2 > 0$$

3. Solve $6x^2 + 4x > 9$

Give your solutions correct to two decimal places.

4. Write the set of integers that satisfy the inequality:
$$7y^2 + 16y < 111$$

5. An inequality is of the form:
$$ax^2 + bx + c \geq 0$$
where a, b and c are integers.

The solution to the inequality is
$$x \leq 1.5 \quad \text{and} \quad x \geq 8$$

Find the values of a, b and c.

READY?

An **arithmetic** (or linear) sequence is one where the terms increase or decrease by the same amount each time: e.g. 1, 7, 13, 19, 25 and 9.8, 9.5, 9.2, 8.9, 8.6
(+6 +6 +6 +6) (−0.3 −0.3 −0.3 −0.3)

The nth term of an arithmetic sequence is of the form $an + b$ where a and b can be any number.

Sequences can also be formed by a multiplicative rule (multiplying by the same number):

2, 4, 8, 16, 32 (×2 ×2 ×2 ×2)

These sequences are called **geometric** sequences. The nth term of a simple geometric sequence is of the form r^n where r can be any number except 0.

e.g. 1 Tick the geometric sequences:

A: 1, 0.1, 0.01, 0.001 ✓ (×0.1 ×0.1 ×0.1)

To check for a geometric sequence, look for a multiplicative rule

B: 12, 36, 144, 720 (×3 ×4 ×5) — This is NOT geometric

C: $\frac{1}{25}$, $\frac{1}{5}$, 1, 5 ✓ (×5 ×5 ×5)

HINT: Work out $b \div a$ if the multiplier to get from a to b is not obvious

D: 252, 21, $\frac{7}{4}$, $\frac{7}{48}$ ✓

$21 \div 252 = \frac{1}{12}$ → ×$\frac{1}{12}$ ×$\frac{1}{12}$ ×$\frac{1}{12}$

e.g. 2 The first three terms of a geometric sequence are

$2\sqrt{3}$, 6, $6\sqrt{3}$

$6 \div 2\sqrt{3} = \sqrt{3}$ → ×$\sqrt{3}$ ×$\sqrt{3}$ This is the term-to-term rule

Find the next two terms of the sequence.

4th term: $6\sqrt{3} \times \sqrt{3} = 18$ 5th term: $18 \times \sqrt{3} = 18\sqrt{3}$

CHECK-IN

1 Find the next three terms:

−13, −5, 3, 11, …

20, 14, 8, 2, …

8, 9, 17, 26, …

2 Find the first four terms of the sequence with nth term $5n − 3$

3 Find the eighth term of the sequence with nth term $n^2 + 12$

e.g. 3 The first term of a geometric sequence is 2. The fourth term of the sequence is −432. Find the fifth term of the sequence.

We can construct an equation (let's use r) to help solve this problem

2, ___, ___, −432   so $2 \times r \times r \times r = -432$
   ×r ×r ×r

$2r^3 = -432$ → $r^3 = -216$ → $r = \sqrt[3]{-216}$ → $r = -6$

So the fifth term is $-432 \times -6 = 2592$

Sequences can be made using different rules. For example, Fibonacci-type sequences are generated by adding two consecutive terms to get the next term.

e.g. 4 A Fibonacci-type sequence is generated by the formula

$$F_{n+2} = F_n + F_{n+1}$$

$F_1 = 3$ and $F_5 = 27$. Work out the value of F_2 and F_4.

Write the first five terms of the sequence using a to represent F_2

3, ___, ___, ___, 27 → 3, a, 3 + a, 2a + 3, 3a + 6

So $27 = 3a + 6$
    $-6 \quad -6$
    $21 = 3a$
    $\div 3 \quad \div 3$
    $7 = a$

$a + 3 + a = 2a + 3$
$3 + a + 2a + 3 = 3a + 6$

So $F_2 = 7$
and $F_4 = 2a + 3 = 2 \times 7 + 3 = 17$

SET?

A. Tick the geometric sequences:

3   12   36   432 ☐

0.96   0.48   0.24   0.12 ☐

1   12   36   144 ☐

0.5   0.15   0.45   0.135 ☐

6   1 $\frac{1}{6}$ $\frac{1}{36}$ ☐

2   $\frac{4}{3}$ $\frac{8}{9}$ $\frac{16}{27}$ ☐

$\frac{1}{2}$ $\frac{1}{3}$ $\frac{1}{5}$ $\frac{1}{8}$ ☐

1   $\sqrt{7}$ 7 $7\sqrt{7}$ ☐

B. The first three terms of a geometric sequence are:

10   $10\sqrt{2}$ 20

Find the next two terms.

C. The first three terms of a geometric sequence are:

0.7   −4.9   34.3

Find the sixth term.

D. Here are the first four terms of a geometric sequence:

$\frac{1}{2}$ x $\frac{9}{32}$ y

Find the value of y.

E. A sequence is generated by the formula $U_{n+2} = U_n + U_{n+1}$

$U_1 = 1.4$ and $U_2 = 3.9$

Work out the value of U_5

F. A sequence is generated by the formula $C_{n+2} = C_n + C_{n+1}$

$C_1 = 6$ and $C_5 = 54$

Work out the value of C_3

GO! SEQUENCES 1

Howay man!

1. S is a geometric sequence.

The third term of S is 120
The sixth term of S is 405

Work out exact value of the first term of the sequence.

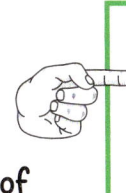

2. A sequence is generated by the formula
$$U_{n+2} = U_n + U_{n+1}$$
$U_1 = 5\sqrt{3}$
$U_6 = 35 + 25\sqrt{3}$

Work out the value of U_2.

3. A population of bats in a cave at the start of year n is B_n. The population at the start of the next year is given by:
$$B_{n+1} = A \times B_n$$
At the start of 2020 the population was 6000. At the start of 2023 the population was 4374.
a) Work out the value of A.
b) Interpret the value of A.

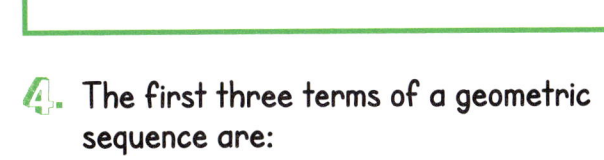

4. The first three terms of a geometric sequence are:
$$(\sqrt{k} - 2) \quad 2 \quad (\sqrt{k} + 2)$$
a) Find the value of k
b) Show that the fifth term of the sequence is
$$2(7 + 5\sqrt{2})$$

READY?

A sequence with n^{th} term of the form $an^2 + bn + c$ (where a, b and c are any number, $a \neq 0$) is called a **quadratic sequence**. For example, the first six terms of the sequence with n^{th} term $n^2 + 5n - 1$ are:

$$5, 13, 23, 35, 49, 65$$

Quadratic sequences have 'second differences' that are equal. First and second differences are found by looking at the additive rule to get from term to term. When the first differences are not equal, we look at the second differences:

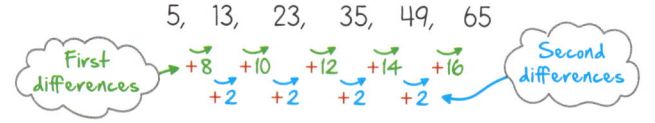

$$5, 13, 23, 35, 49, 65$$

First differences: +8, +10, +12, +14, +16
Second differences: +2, +2, +2, +2

e.g. 1 Which of these sequences has a quadratic n^{th} term?

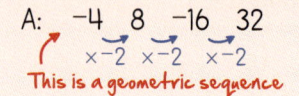

A: −4, 8, −16, 32 (×−2, ×−2, ×−2) — This is a geometric sequence

B: 8, 15, 22, 29 (+7, +7, +7) — This is an arithmetic sequence

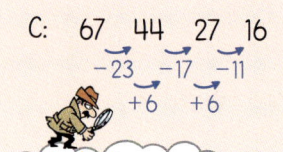

C: 67, 44, 27, 16 (−23, −17, −11; +6, +6) — Look... the second differences are equal

D: 10, 12, 22, 34 (+2, +10, +12; +8, +2) — This is not a quadratic sequence (it's Fibonacci-type)

Sequence C is quadratic

e.g. 2 The nth term a sequence is $2n^2 − 5n + 3$
Find the 20th term of the sequence.
$n = 20$ so $2 \times 20^2 − 5 \times 20 + 3 = 703$

TECHNICAL STUFF: $2n^2 − 5n + 3 = 2n^2 + {-5n} + 3$

CHECK-IN

1. Find the nth term:
   27, 31, 35, 39, 43, ...
   −9, −7, −5, −3, −1, ...
   10, 7, 4, 1, −2, ...
   13.5, 18, 22.5, 27, 31.5, ...

The second differences also help to find the nth term.

e.g. 3 Find the nth term of the quadratic sequence:

1. Find the second difference:
   12, 27, 50, 81, 120
   +15, +23, +31, +39
   +8, +8, +8

2. Halve the second difference to find the coefficient of n^2:
   $8 \div 2 = 4$ so the nth term starts with $4n^2$

3. Compare the sequence with the first few terms of $4n^2$
   $4n^2 \to$ 4, 16, 36, 64, 100
   +8, +11, +14, +17, +20
   sequence → 12, 27, 50, 81, 120

4. Find the nth term of the adjustment:
   The nth term of 8, 11, 14, 17, 20 is $3n + 5$

5. Construct the nth term of the sequence: $4n^2 + 3n + 5$

e.g. 4 The first five terms of a quadratic sequence are:
$$3, 5, 11, 21, 35$$
Find the nth term of the sequence.

1. 3, 5, 11, 21, 35
   +2, +6, +10, +14
   +4, +4, +4

2. $4 \div 2 = 2$ so the nth term starts with $2n^2$

3. Compare the sequence with the first few terms of $2n^2$
   $2n^2 \to$ 2, 8, 18, 32, 50
   +1, −3, −7, −11, −15
   sequence → 3, 5, 11, 21, 35

4. The nth term of 1, −3, −7, −11, −15 is $−4n + 5$

5. The nth term is $2n^2 − 4n + 5$

A. Tick the quadratic sequences:

7, 13, 21, 31, 43 ☐

9, 18, 25, 34, 41 ☐

3, −2, −11, −24, −41 ☐

1200, 900, 675, 506.25 ☐

B. The nth term of a sequence is:
$$3n^2 + 4n − 1$$
Find the 15th term of the sequence.

C. Find the nth term of these quadratic sequences:
(i) 7, 24, 51, 88, 135, ...

(ii) 7, 20, 37, 58, 83, ...

(iii) 9, 19, 35, 57, 85, ...

D. Find the nth term of these quadratic sequences:
(i) 3, 4, 7, 12, 19, ...

(ii) 7, 8, 13, 22, 35, ...

SET?

GO! SEQUENCES 2

Glorious!

1. Find the nth term of these quadratic sequences:

a) 0, −3, −4, −3, 0, ...

b) $\frac{11}{4}$, 3, $\frac{15}{4}$, 5, $\frac{27}{4}$, ...

2. Q is a quadratic sequence.

The first term of Q is 4
The third term of Q is 10
The fifth term of Q is 32

Find the nth term of the sequence.

3. Place the digits 1, 2, 3, 4 and 5 to complete the statement.

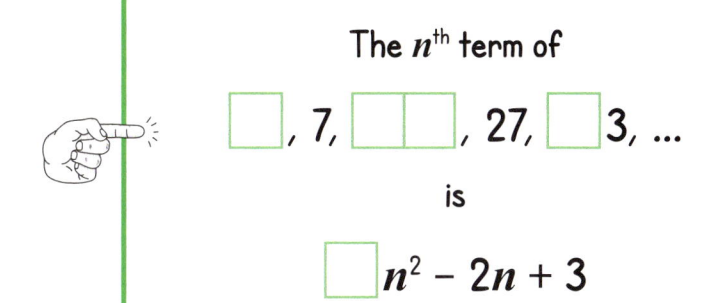

The nth term of

☐, 7, ☐☐, 27, ☐3, ...

is

☐$n^2 - 2n + 3$

4. The first five terms of a quadratic sequence are:

$5a$, $6a + b$, $7a + 4b$, $8a + 9b$, $9a + 16b$

Show that the nth term of the sequence is

$b(n-1)^2 + a(n+4)$

READY?

If two variables are connected by a multiplicative relationship, then the variables are in **direct proportion**. For example, in August 2024 the exchange rate between pounds (GBP) and euros (EUR) was 1 GBP = 1.18 EUR. To convert any amount in pounds to an amount in euros we would multiply by 1.18. This number can be called the multiplier.

In general, we can use algebra to describe these relationships. The symbol '$\propto$' is used to indicate that a directly proportional relationship exists. So, if y is directly proportional to x:

$$y \propto x \quad \text{and} \quad y = kx$$

k is the multiplier

Using this notation can help us to solve more complex proportion problems.

CHECK-IN

1 y is directly proportional to x. Find the missing values:

x	5	8
y	35	

x	6	10
y	9	

x	12	7
y		1.75

x	21	
y	49	35

2 In which of these graphs is y directly proportional to x?

Graph A, Graph B, Graph C

e.g. 1 y is directly proportional to x.
When $x = 72$, $y = 63$

a) Find a formula connecting x and y

1 Write a mathematical statement summarising the relationship:
$$y \propto x \quad \text{so} \quad y = kx$$

2 Use the known values of x and y to work out the value of k:
When $x = 72$ and $y = 63$: $y = kx \rightarrow 63 = k \times 72$
$\div 72 \quad \div 72$

We could write this as $k = \frac{7}{8}$ $\rightarrow 0.875 = k$

3 Write the formula: $y = 0.875x$

b) Work out the value of x when $y = 119$

$y = 0.875x \rightarrow 119 = 0.875x$
$\div 0.875 \quad \div 0.875$
$136 = x \quad \text{or} \quad x = 136$

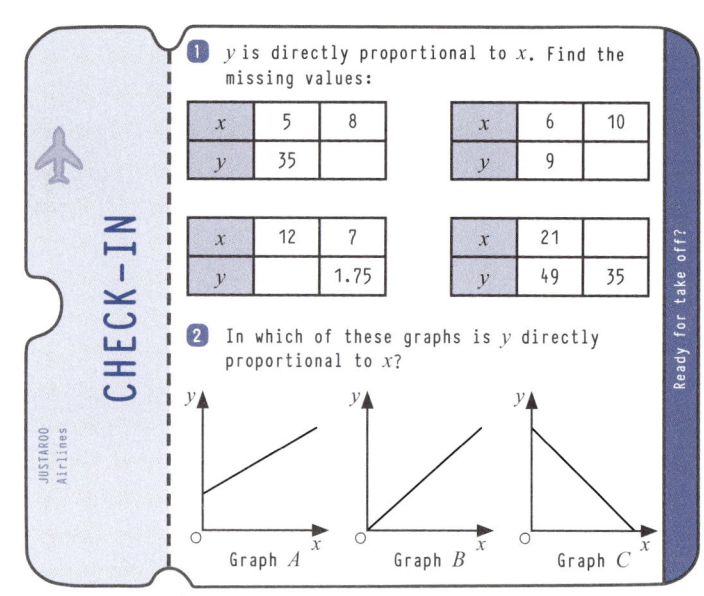

Sometimes there are further operations that need to be considered ...

e.g. 2 t is directly proportional to the square of s.
When $s = 4$, $t = 104$

a) Work out the value of t when $s = 7$

Start by finding a formula connecting s and t

1 $t \propto s^2$ so $t = ks^2$

2 Work out the value of k:
When $s = 4$ and $t = 104$: $t = ks^2$
$\rightarrow 104 = k \times 4^2$
$\rightarrow 104 = k \times 16$
$\div 16 \quad \div 16$
$6.5 = k$ so $t = 6.5s^2$ **3**

Now... answer the question

When $s = 7$: $t = 6.5 \times 7^2 = 318.5$

b) Given that $s > 0$, find the value of s when $t = 936$

$t = 6.5s^2 \rightarrow 936 = 6.5s^2$
$\div 6.5 \quad \div 6.5$
$144 = s^2$ so $s = 12$

SET?

A. y is directly proportional to x.
Find a formula connecting x and y if:

(i) $y = 42$ when $x = 14$

(ii) $y = 23$ when $x = 92$

(iii) $y = 78$ when $x = 15$

B. T is directly proportional to n.
When $n = 9$, $T = 121.5$
Find the value of T when $n = 20$

C. y is directly proportional to x.
When $x = 85$, $y = 136$
Find the value of x when $y = 184$

D. y is directly proportional to the square of x.
When $x = 3.5$, $y = 294$
Find the value of y when $x = 1.2$

E. w is directly proportional to v^3.
$w = -320$ when $v = -4$
Find v when $w = 1715$

F. y is directly proportional to $\sqrt{x}$.
$y = 4.34$ when $x = 1.96$
Find x when $y = 5.89$

GO! DIRECT PROPORTION — Cracking!

1. y is directly proportional to x^2.
   When $x = 2.5$, the value of $y = 250$
   Find the value of y when $x = 5$

   Cedric says,

   "2.5 is doubled to get 5, so the answer is $2 \times 250 = 500$"

   Do you agree with Cedric? Explain why.

2. A stone is dropped into a well.

   The speed of the stone is directly proportion to the square of the time that the stone has been falling.

   After two seconds the stone is travelling at 19.6 metres per second. The stone hits the bottom of the well after three seconds. What is the speed of the stone at the moment of impact?

3. y is directly proportional to x^2.
   x is directly proportional to w.
   $y = 9072$ when $w = 9$

   Find the value of y when $w = 8$

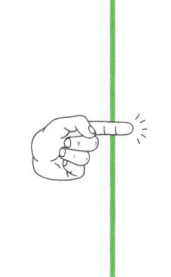

   HINT: Substitute one formula into the other

4. A pendulum has length L.

   The time, T, that the pendulum takes to complete a full swing is directly proportional to the square root of L.

   L is reduced by 19%. What is the percentage reduction in T?

READY?

If one variable increases at the same rate as another decreases, then the variables are **inversely proportional**. For example, if there are 200 sandwiches available at a buffet lunch, then the number of people is inversely proportional to the number of sandwiches they could have each (if shared equally!)

People	Sandwiches
4	50
8	25

×2 ↓ ↓ ÷2

Notice that 4 × 50 = 200
... and also 8 × 25 = 200

If y is inversely proportional to x, then it is also true that y is directly proportional to $\frac{1}{x}$. Therefore:

$$y \propto \frac{1}{x} \quad \text{and} \quad y = \frac{k}{x}$$

CHECK-IN

1 y is inversely proportional to x. Find the missing values:

x	3	12
y	40	

x	25	
y	24	30

x	50	7.5
y		20

x		20
y	6.4	4.8

2 In which of these graphs is y inversely proportional to x?

Graph A   Graph B   Graph C

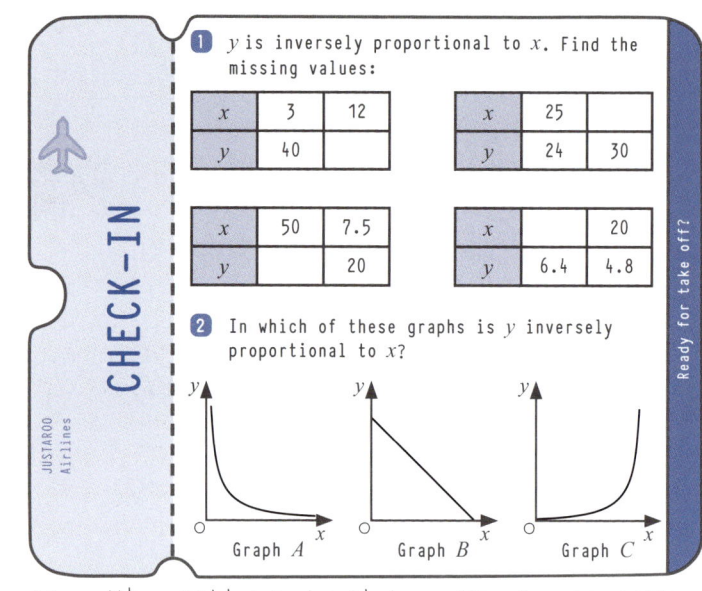

As with problems involving direct proportion, sometimes the relationship arises when an operation is applied to one of the variables ...

e.g. 1 y is inversely proportional to x.
When $x = 20$, $y = 4$
a) Find a formula connecting x and y

1 Write a mathematical statement summarising the relationship:
$$y \propto \frac{1}{x} \quad \text{so} \quad y = \frac{k}{x}$$

2 Use the known values of x and y to work out the value of k:
When $x = 20$ and $y = 4$:
$y = \frac{k}{x} \rightarrow 4 = \frac{k}{20}$ so $80 = k$

3 Write the formula: $y = \frac{80}{x}$

b) Work out the value of x when $y = 16$
$y = \frac{80}{x} \rightarrow 16 = \frac{80}{x} \rightarrow 16x = 80 \rightarrow x = 5$
×x ×x ÷16 ÷16

e.g. 2 y is inversely proportional to $\sqrt{x}$.
When $x = 9$, $y = 8$
a) Work out the value of y when $x = 144$

First find a formula connecting x and y:
1 $y \propto \frac{1}{\sqrt{x}}$ so $y = \frac{k}{\sqrt{x}}$

2 When $x = 9$ and $y = 8$: $8 = \frac{k}{\sqrt{9}} \rightarrow 8 = \frac{k}{3} \rightarrow 24 = k$

3 so $y = \frac{24}{\sqrt{x}}$

Therefore, when $x = 144$: $y = \frac{24}{\sqrt{144}} = \frac{24}{12} = 2$

b) Work out the value of x when $y = 10$
$y = \frac{24}{\sqrt{x}} \rightarrow 10 = \frac{24}{\sqrt{x}} \rightarrow 10\sqrt{x} = 24 \rightarrow \sqrt{x} = 2.4$
×√x ×√x ÷10 ÷10 □² □²

so $x = 5.76$

A. y is inversely proportional to x.
When $x = 24$, $y = 2.5$
(i) Find a formula connecting x and y.
(ii) Find the value of y when $x = 40$

B. p is inversely proportional to q.
When $p = 18$, $q = 16$
(i) Find a formula connecting p and q.
(ii) Find the value of p when $q = 6$

C. y is inversely proportional to the cube of x.
When $x = 10$, $y = 0.008$
Find the value of y when $x = 4$

D. y is inversely proportional to the square root of x.
When $x = 16$, $y = 5$
Find the value of x when $y = 2$

SET?

E. s is inversely proportional to t^2.
$s = 80$ when $t = 0.75$
Find the positive value of t when $s = 125$

F. y is inversely proportional to $\sqrt{x}$.
$y = 7.2$ when $x = 0.04$
Find x when $y = 1.8$

GO!

INVERSE PROPORTION

Spot on!

1. It is thought that it took 15 000 people six years to build Hadrian's Wall in the north of England. Work began in the year 122.
   a) If there had only been 10 000 people building the wall, when would the work have been finished?
   b) Describe an assumption you have made in your working for part (a).

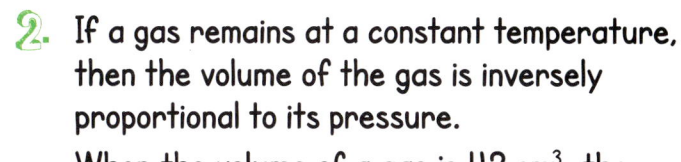

2. If a gas remains at a constant temperature, then the volume of the gas is inversely proportional to its pressure.

   When the volume of a gas is 42 cm^3, the pressure is 80 000 pascals.

   The temperature of the gas remains the same and the volume of gas is increased by 100%. Calculate the percentage change in pressure.

3. a is directly proportional to b^2.
   b is inversely proportional to c.

   When $a = 1.8$, $c = 10$

   Find the positive value of c when $a = \dfrac{5}{4}$

   HINT: Substitute one formula into the other

4. y is inversely proportional to $\sqrt{x}$.
   x is directly proportional to w^3.

   When $w = 4$, $y = 0.375$

   Find the value of y when $w = 9$

READY?

When a **shape** is **enlarged** by a scale factor, all of the side lengths are multiplied by that number. Drawing **rays** through the **centre of enlargement** and matching vertices, we can see that **negative scale factors** can be used as well:

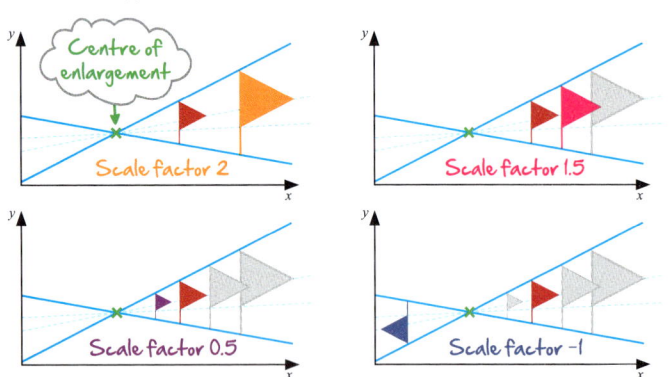

- Centre of enlargement
- Scale factor 2
- Scale factor 1.5
- Scale factor 0.5
- Scale factor −1

CHECK-IN

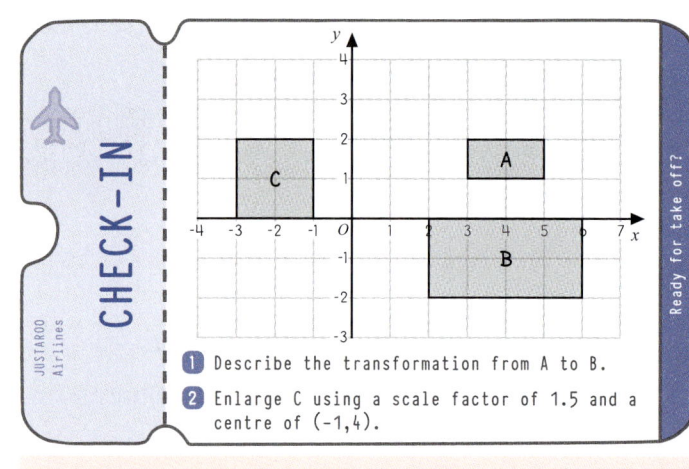

1. Describe the transformation from A to B.
2. Enlarge C using a scale factor of 1.5 and a centre of (−1,4).

e.g. 1 Describe fully the single transformation that maps shape A onto shape B.

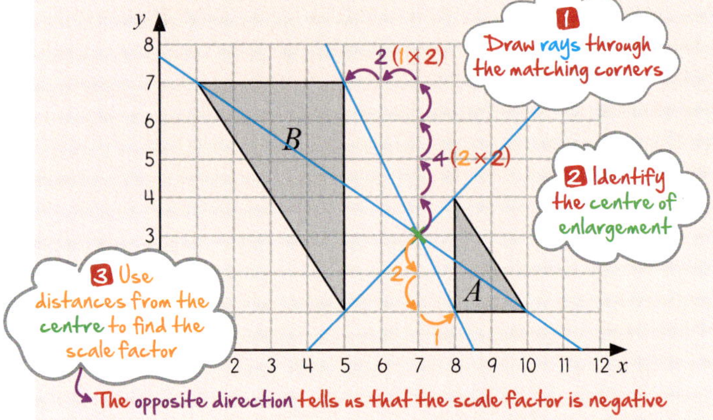

1. Draw rays through the matching corners
2. Identify the centre of enlargement
3. Use distances from the centre to find the scale factor

The opposite direction tells us that the scale factor is negative

A is mapped onto B by an enlargement using the centre (7,3) and a scale factor of −2

e.g. 2 Enlarge the shape using a scale factor of −1.5 and (4,3) as the centre of enlargement.

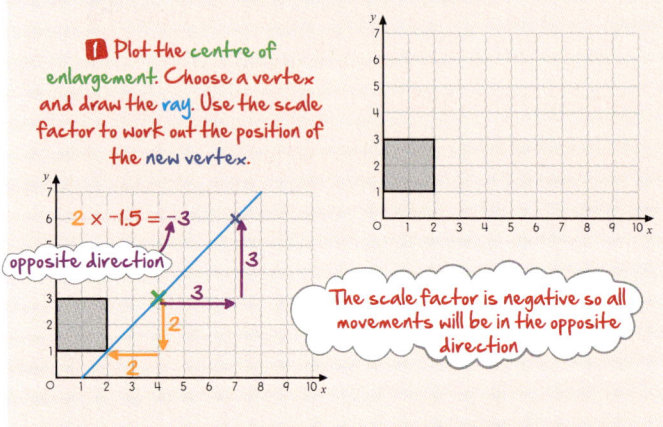

1. Plot the centre of enlargement. Choose a vertex and draw the ray. Use the scale factor to work out the position of the new vertex.

   $2 \times -1.5 = -3$

   opposite direction

   The scale factor is negative so all movements will be in the opposite direction

2. Choose another vertex and repeat. The two new vertices can be joined.

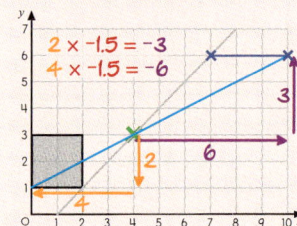

   $2 \times -1.5 = -3$
   $4 \times -1.5 = -6$

3. Repeat for all vertices and join them up to create the new shape.

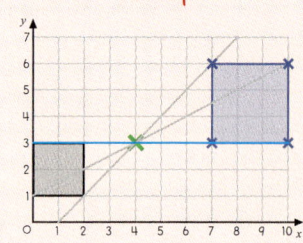

SET?

A. Enlarge shape A using a scale factor of −2 and (5,7) as the centre of enlargement.

B. Enlarge shape B using a scale factor of −3 and (4,12) as the centre of enlargement.

C. Enlarge shape C using a scale factor of −0.5 and (13,4) as the centre of enlargement.

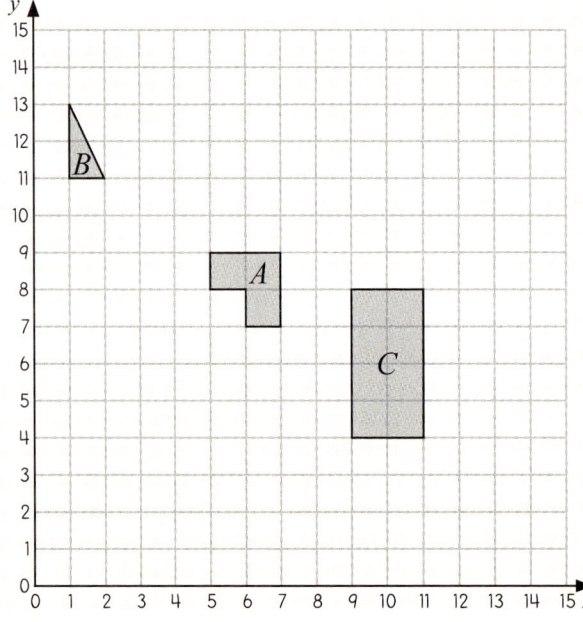

D. Describe the transformation that maps shape D onto shape E.

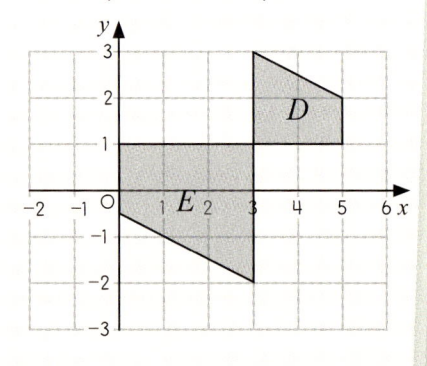

GO! ENLARGEMENTS

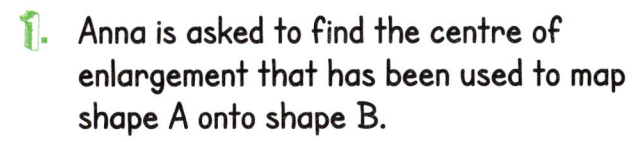

Cool!

1. Anna is asked to find the centre of enlargement that has been used to map shape A onto shape B.

   Her rays are shown on the diagram.

   Anna thinks it is not possible to find the centre of enlargement unless the grid is extended.

   Do you agree? Explain why.

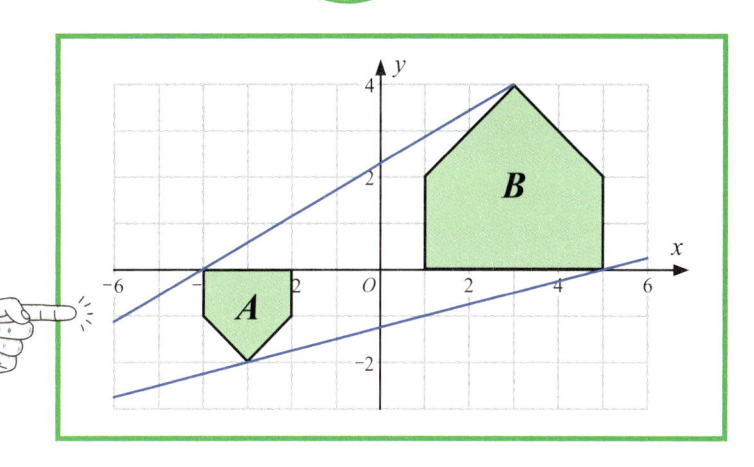

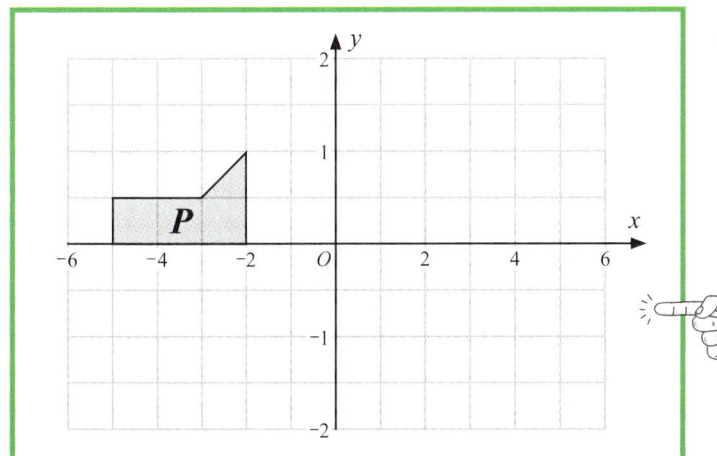

2. Enlarge shape P using a scale factor of -2 and $(-2, 0.5)$ as the centre of enlargement. Label the image Q.

3. $ABCDE$ is mapped onto $PQRST$ by an enlargement:

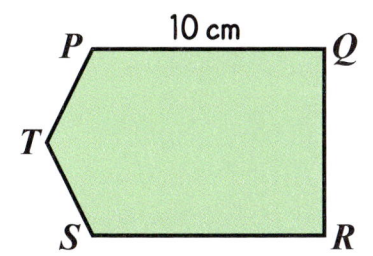

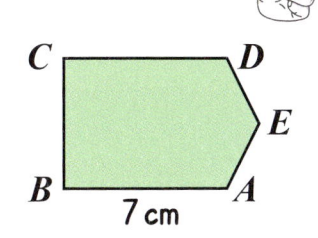

   10 cm

   7 cm

   State the exact value of the scale factor of the enlargement.

4. An enlargement with scale factor -4 is carried out using $(-1, 11)$ as the centre of enlargement.

   Work out where the point $(8, 23)$ is moved to by the enlargement.

READY?

Translations, rotations and reflections are the types of **transformation** that always result in a congruent shape (shapes that are the same shape and same size). A combination of them can often be described as a SINGLE transformation.

e.g. 1 Shape A is mapped onto shape B by a rotation of 180° about the point $(-1, 2)$.

Shape B is mapped onto C by a translation of $\begin{pmatrix} 2 \\ -4 \end{pmatrix}$

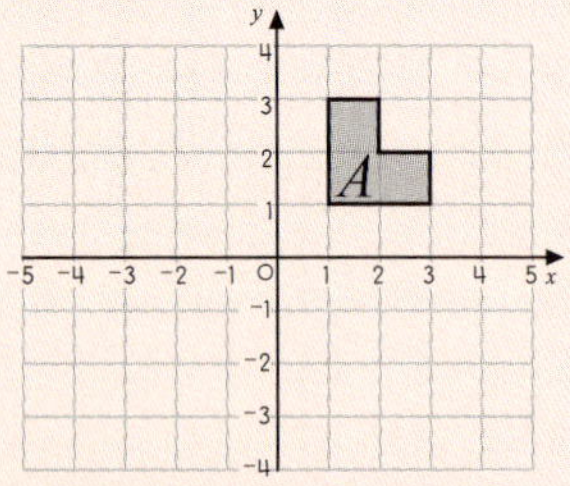

Describe fully the single transformation that maps shape A onto shape C.

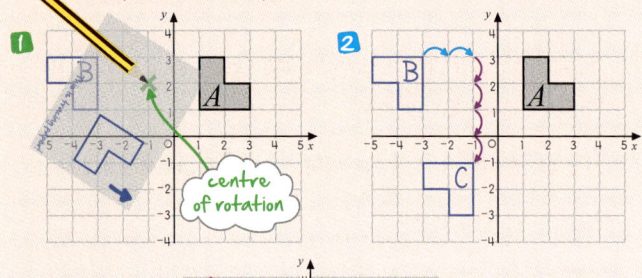

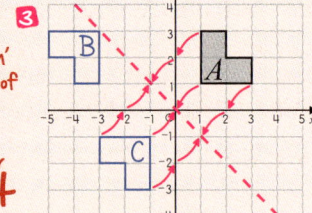

A is mapped onto C by a reflection in the line $y = -x$

'Single transformation' means just one of reflection, rotation, translation or enlargement

CHECK-IN

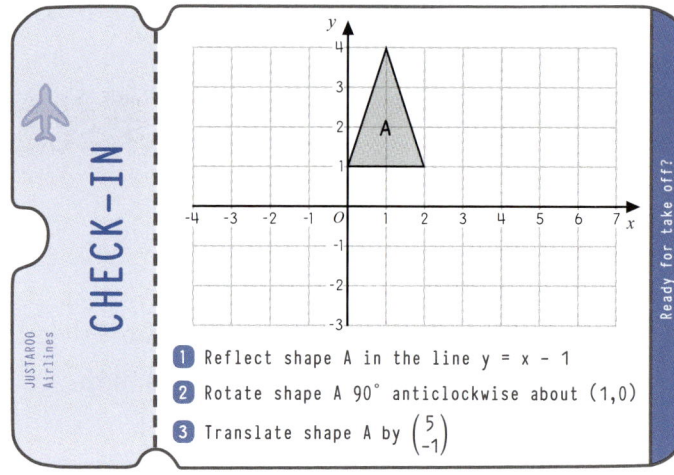

1. Reflect shape A in the line $y = x - 1$
2. Rotate shape A 90° anticlockwise about $(1, 0)$
3. Translate shape A by $\begin{pmatrix} 5 \\ -1 \end{pmatrix}$

Sometimes, a point on (or in) a shape remains in the same position when the shape is transformed. Any of these points are called **invariant points**.

e.g. 2 Shape P is reflected in the line $y = x - 1$ and the image is labelled Q.

Write the coordinates of any invariant points.

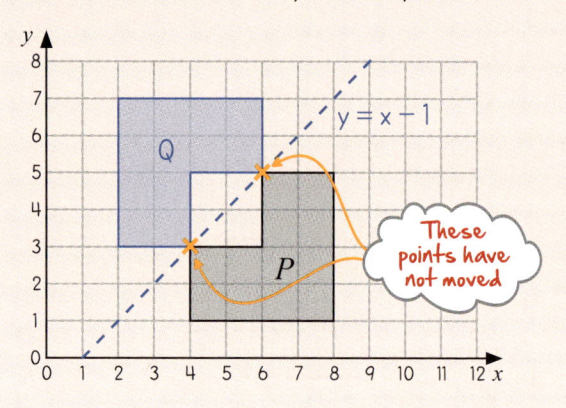

These points have not moved

There are invariant points at $(4, 3)$ and $(6, 5)$

Vertices in the same position do NOT always indicate invariant points, e.g. this rotation has only one invariant point, not these vertices

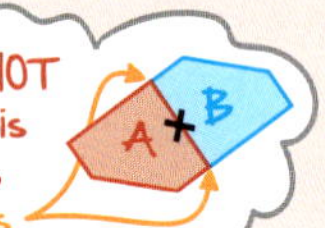

SET

A. Translate A by $\begin{pmatrix} 2 \\ -2 \end{pmatrix}$ and label the image L. Rotate L 180° about the point $(7, 11)$. Label the image W. Describe the single transformation that maps A onto W.

B. Reflect B in the line $x + y = 10$ and label the image M. Reflect M in the line $x = 5$. Label the image X. Describe the single transformation that maps B onto X.

C. Reflect C in the line $x = 4$ and label the image N. Reflect N in the line $y = x$. Label the image Y. Describe the single transformation that maps C onto Y.

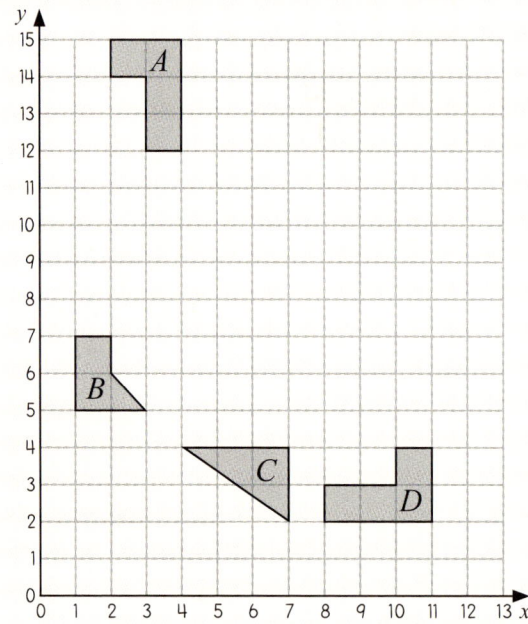

D. Rotate D 270° clockwise about the point $(9, 6)$. Label the image P. Rotate P 270° anticlockwise about the point $(10, 8)$. Label the image Z. Describe the single transformation that maps D onto Z.

E. Describe any vertices that are invariant points under the transformations in questions A to D.

GO! TRANSFORMATIONS

skills!

1. Shape A is rotated 180° about the point $(0,1)$ and labelled B. The image is then translated by $\begin{pmatrix} a \\ b \end{pmatrix}$ and labelled C.

   A can also be mapped onto C by a rotation of 180° about the point $(-1, 0)$.
   Work out the values of a and b.

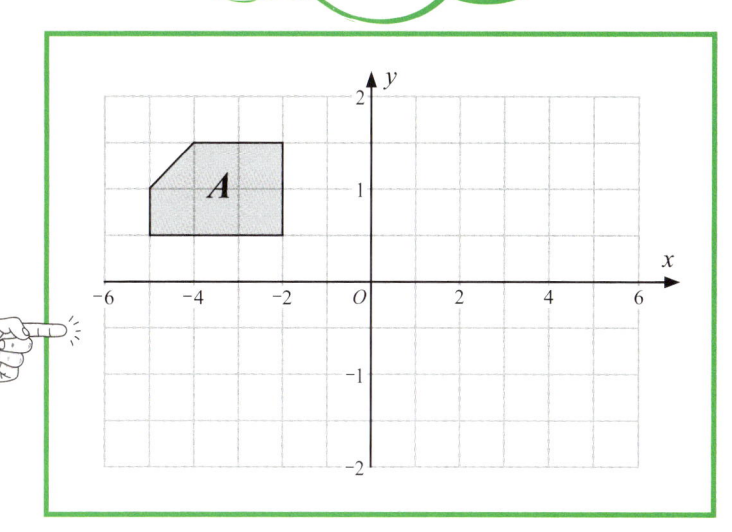

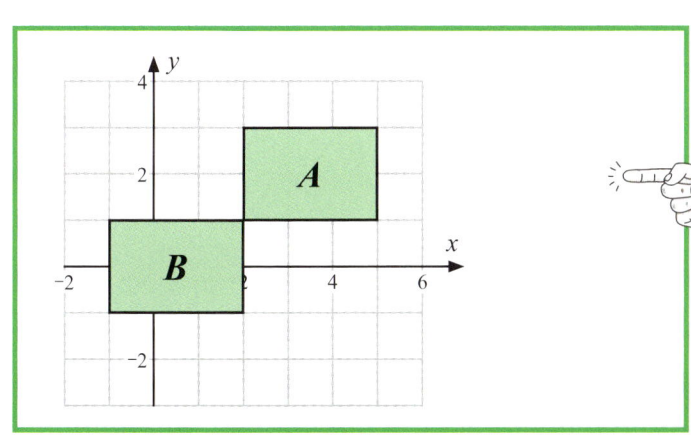

2. State the number of invariant points if A is mapped onto B by:

   a) A reflection in the line $y = 1$ followed by a translation by $\begin{pmatrix} -3 \\ 0 \end{pmatrix}$

   b) A reflection in the line $x = 2$ followed by reflection in the line $y = 1$

   c) A 90° clockwise rotation about $(1, 3)$ followed by a reflection in the line $y = x$

3. Shape P is mapped onto shape Q by a rotation of 180° about the point $(0, 1)$.

   Shape Q is mapped onto shape R by a translation using the vector $\begin{pmatrix} 4 \\ 0.5 \end{pmatrix}$.

   Shane thinks that there are two invariant points in the single transformation that maps shape P onto shape R.

   Explain why Shane is not correct.

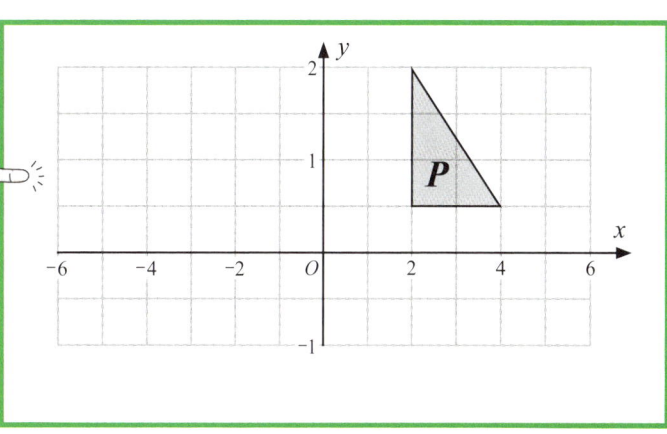

4. Decide whether the following statement is always true, sometimes true or never true:

   An enlargement with scale factor -3 has at least one invariant point

   Explain your reasoning.

READY?

By studying the angles that are made by joining points on the circumference of a circle we see some patterns. These patterns are known as **circle theorems**.

For example, if O is the centre of a circle and A, B and C are points on the circumference, then angle AOB is always double the size of angle ACB. This is sometimes described as:

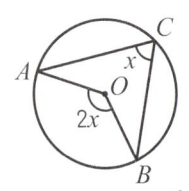

The angle at the centre is double the angle at the circumference

We can use this fact to find missing angles …

e.g. 1
A, B and C are points on a circle with centre O. The reflex angle $AOB = 200°$. Calculate the size of angle ACB.

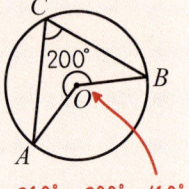

1 Find any missing angles using known facts

Angle AOB = 160° because angles meeting at a point add up to 360°

$360° - 200° = 160°$

We label all the angles that we find

2 Use the circle theorem to find the required angle

Angle ACB = 80° because the angle at the centre is double the angle at the circumference

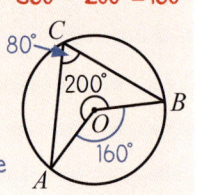

If we imagine moving the point C around the circle between A and B, angle ACB must always stay the same size. We say:

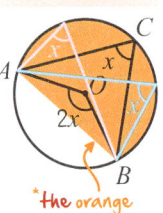

Angles in the same segment* are equal

*the orange segment

CHECK-IN

1. Label the circle using words from the list:
   - circumference
   - radius
   - diameter
   - chord

2. Has this circle been split into two sectors or two segments?

e.g. 2
P, Q, R and S are points on a circle. Angle $PRQ = 35°$. Work out the size of angle PSQ.

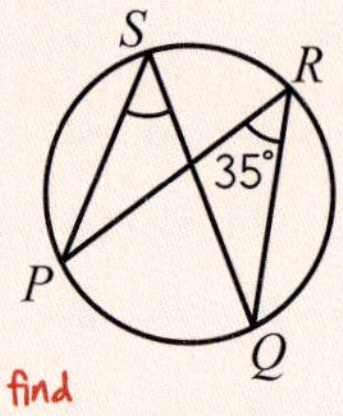

1 Draw a chord to identify a segment

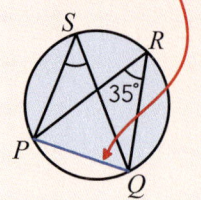

2 Use the circle theorem to find the required angle

Angle PSQ = 35° because angles in the same segment are equal

If point A is moved so that AB is a diameter, then angle ACB must be a right angle. This is sometimes stated as:

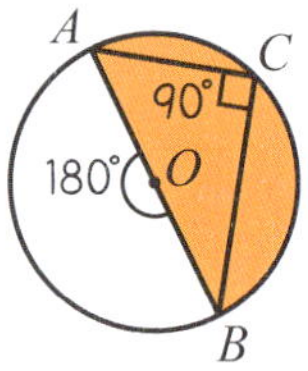

The angle in a semicircle is a right angle

We can prove that the circle theorems are true …

e.g. 3
AB is a diameter of a circle with centre O. Prove that angle ACB is a right angle.

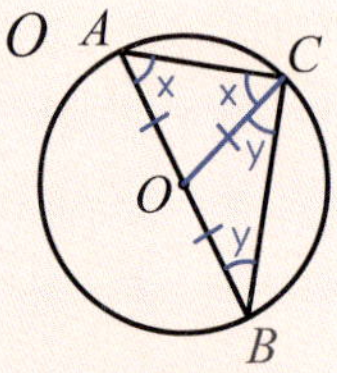

1 Draw a radius and look for isosceles triangles

Triangles AOC and BOC are isosceles

2 Create a chain of logical steps

OA = OC so angles OAC and OCA are equal … Label the angles x
OB = OC so angles OBC and OCB are equal … Label the angles y
From triangle ABC, $x + x + y + y = 2x + 2y = 180°$
So, $x + y = 90°$ … which means that angle ACB is a right angle

SET?

A. Find the value of the labelled angles. Give reasons for your answers.

(i) circle with S, R, P, Q; angle $57°$ at S, angle $x°$ at R

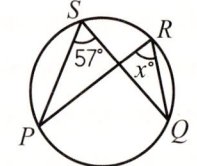

(ii) circle with centre O; points M, N, L; angle $y°$, $132°$

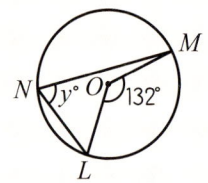

(iii) circle with A, B, C, centre O; $78°$, $a°$

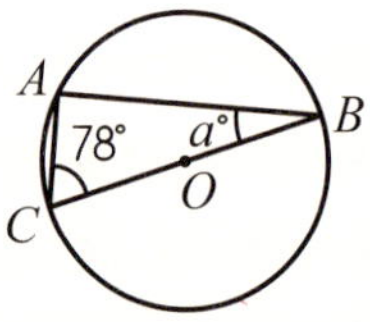

(iv) circle with A, B, C, D; $100°$, $b°$, $60°$

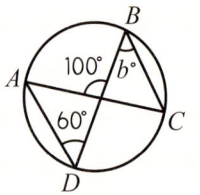

(v) circle with Q, R, P, S; $35°$, $48°$, $41°$, $p°$

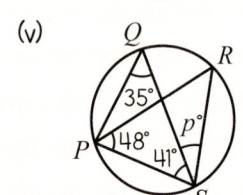

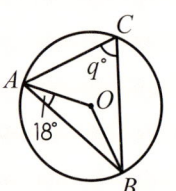

(vi) circle with A, C, B, centre O; $18°$, $q°$

B. Work out the size of angle MNJ. Give reasons for your answer.

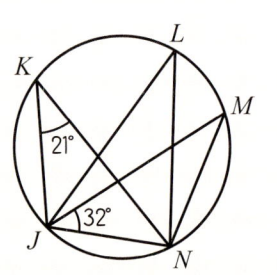

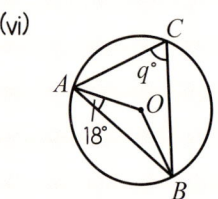

circle with K, L, M, J, N; $21°$, $32°$

CIRCLE THEOREMS 1

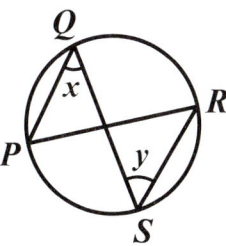

1. Peter thinks that the angles labelled x and y are equal.

Do you agree? Explain why.

2. Find the value of x.

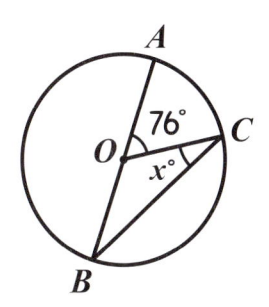

Justify your answer by finding a second way to work out this angle.

3. Work out the size of the angle labelled a.

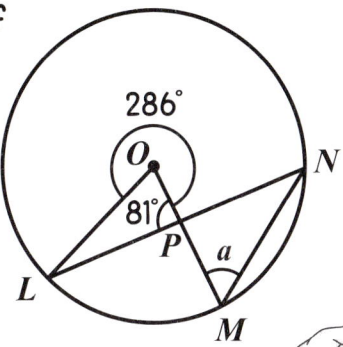

State any circle theorems you use.

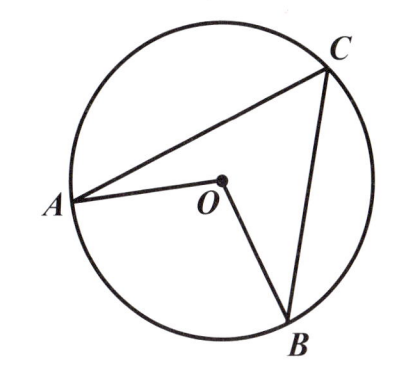

4. Prove that angle $AOB = 2 \times$ angle ACB

READY?

More **circle theorems** can be found by studying chords and tangents to circles. For example …

A and B are points on a circle. The tangents from A and B meet at a point outside the circle. Call this point P. Then it can be shown that
$$AP = BP$$ 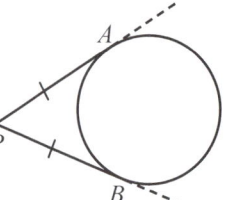
We say:

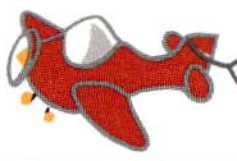

Two tangents that meet at a point are equal in length

e.g. 1 XZ and YZ are tangents to a circle.
$XZ = 13$ cm
Work out the length YZ.
Give a reason for your answer.

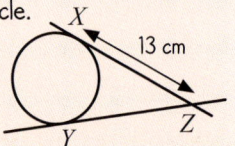

$YZ = 13$ cm because tangents that meet at a point are equal in length

If O is the centre of the circle, then the radius OA makes a right angle with the tangent at A. We say:

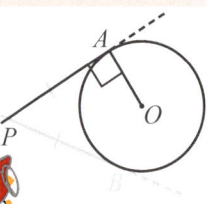

A radius and tangent that meet are perpendicular

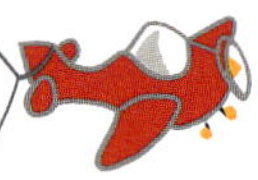

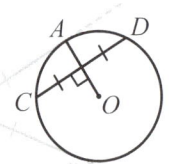

If C and D are points on the circle, and the chord CD is perpendicular to the radius OA, then the radius bisects the chord. We often describe this as:

The radius that is perpendicular to a chord also bisects the chord

CHECK-IN

1 Complete the sentences:

A _____ is a straight line outside a circle that touches the circle exactly once

A _____ splits a circle into two segments

2 Calculate the length AC in these triangles:

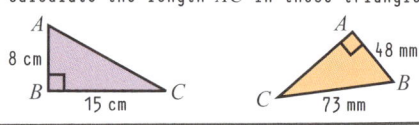

(triangle 1: $AB = 8$ cm, $BC = 15$ cm, right angle at B)
(triangle 2: $AB = 48$ mm, $BC = 73$ mm, right angle at B)

e.g. 2 A and B are points on a circle with centre O.
MN is a tangent to the circle at A.
Angle $OBA = 31°$
Calculate the value of y.

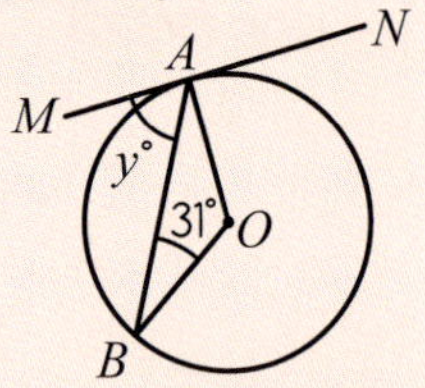

1 Find any missing angles using known facts

OA and OB are both radii so triangle AOB is isosceles and angle OAB = angle OBA
Therefore angle OAB = 31°

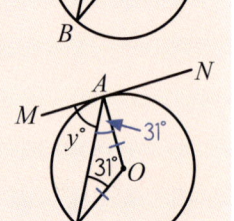

2 Identify and use a circle theorem to find the required angle

Angle OAM = 90° because a radius and tangent that meet are perpendicular
Therefore angle BAM = 90° − 31° = 59°

So $y = 59$

e.g. 3 A, B and C are points on a circle with centre O.
OA is a radius of the circle.
The chord BC is perpendicular to OA.
$BP = 4$ cm and $AC = 5$ cm.
Calculate the length AP.

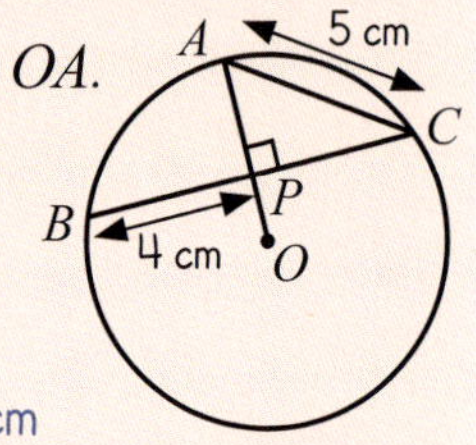

1 Identify and use a circle theorem to find a missing length

The radius bisects the chord so CP = 4 cm

2 Identify other known facts that can help find a missing length

Triangle ACP is right-angled, so use Pythagoras' theorem:
$AP^2 + CP^2 = AC^2$ so $AP^2 + 4^2 = 5^2$ so $AP^2 + 16 = 25$
Therefore $AP^2 = 9$ so $AP = 3$ cm

A. Work out the missing values. Justify your answers.

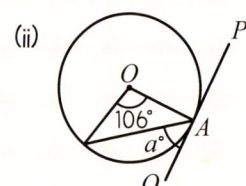

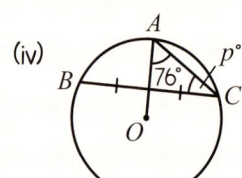

(i) angle $26°$ at O, $x°$ marked
(ii) $106°$ at O, $a°$ marked
(iii) 6 cm, y cm
(iv) $76°$, $p°$
(v) 19 cm, b cm
(vi) $q°$, $72°$

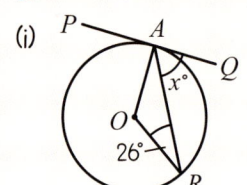

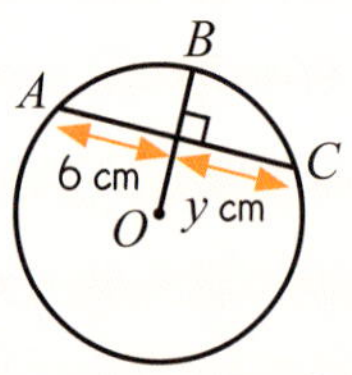

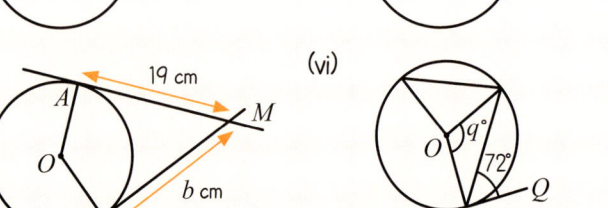

B. Calculate the missing values.

(i) $AB = 6$ cm, $BC = 10$ cm, v cm
(ii) $PQ = 24$ cm, 7 cm, t cm

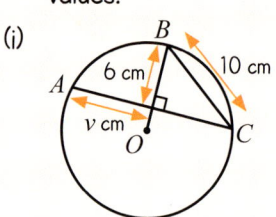

SET?

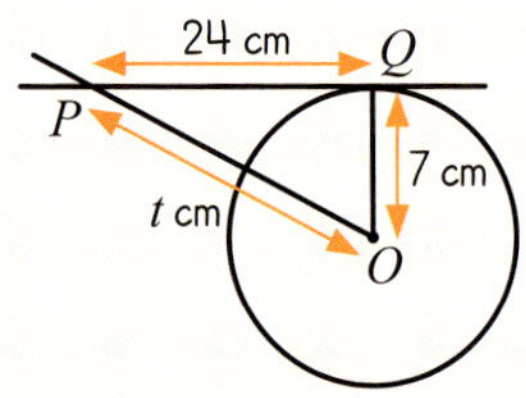

C. Work out the length OC.

$OA = 5$ cm, $AC = 6$ cm

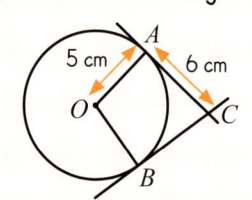

Give your answer to two significant figures.

GO! CIRCLE THEOREMS 2

Invincible!

1. Maddie thinks that the lengths QR and SR are equal.

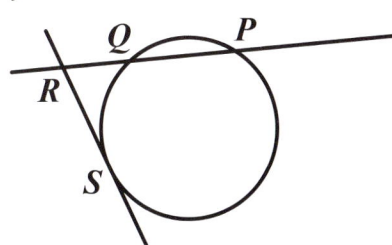

    Do you agree? Explain why.

2. A, B and C are points on the circle. Find the value of y.

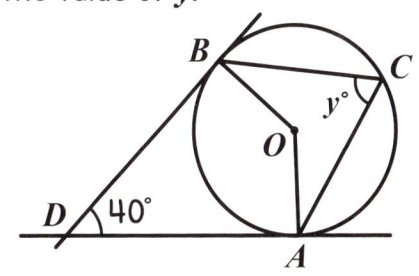

    Give reasons for your answer.

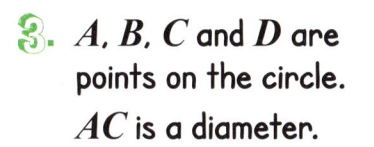

3. A, B, C and D are points on the circle. AC is a diameter.

    Work out the size of angle labelled x.

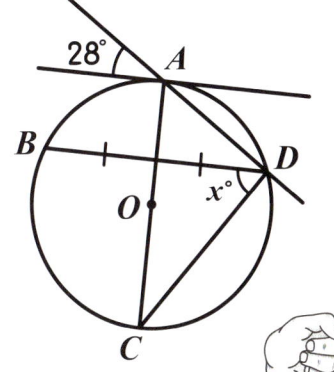

    Justify your answer.

4. BC is a diameter of the circle. LM is a tangent to the circle at the point B.

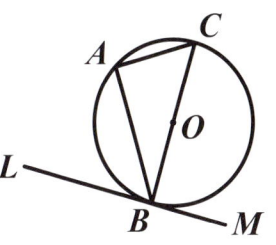

    Prove that:
      Angle ABL = Angle ACB.

READY?

We already know that angles in the same segment* are equal.

If we draw a tangent at the point B we can work out another **circle theorem**...

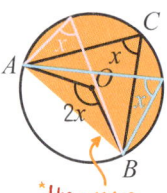

*the orange segment

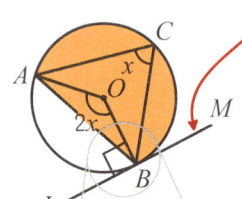

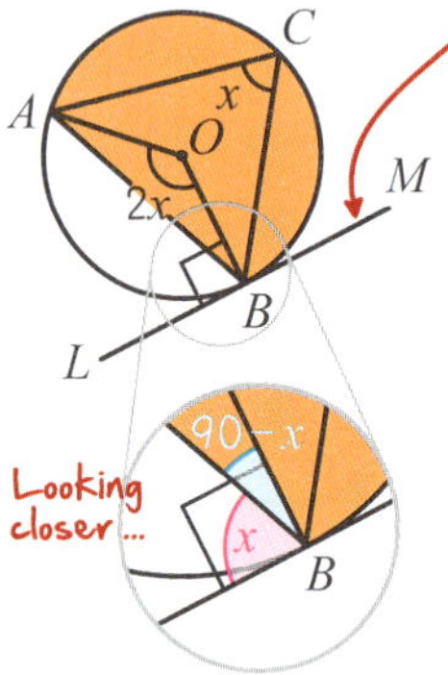

Triangle AOB is isosceles, so angles OAB and OBA are equal:

So angle $OBA = \dfrac{180 - 2x}{2} = 90 - x$

A radius and tangent at a point are perpendicular:

So angle $ABL = 90 - (90 - x) = x$

Therefore angle $ABL =$ angle ACB

We call this: **The alternate segment theorem**

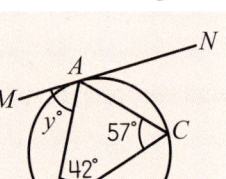

e.g. 1 A, B and C are points on a circle. MN is a tangent at the point A.

Angle $ACB = 57°$
Angle $ABC = 42°$

State the value of y. Give a reason for your answer.

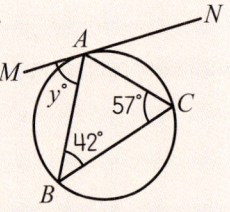

1 Use a chord to identify the **alternate segment** (the one opposite y)

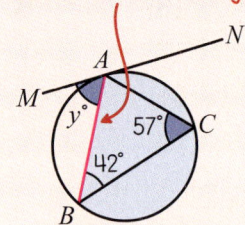

2 Use the circle theorem to find the required angle

$y = 57$ because of the alternate segment theorem

Simply stating the theorem as a reason is fine

CHECK-IN

1 Find the size of the angle labelled x:

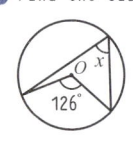

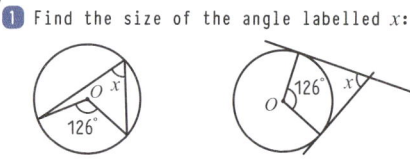

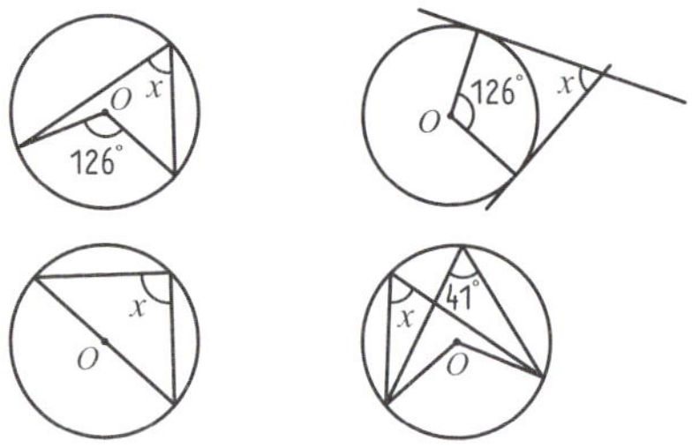

If we join any four points on the circumference of a circle we create a **cyclic quadrilateral**. It can be seen that:

$a + c = 180°$ and $b + d = 180°$

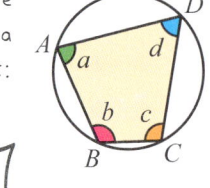

The opposite angles in a cyclic quadrilateral add up to 180°

e.g. 2 $PQRS$ is a cyclic quadrilateral

Angle $QPS = 81°$
Angle $PQR = 88°$

Work out the value of x.

$x = 92$ because opposite angles in a cyclic quadrilateral add up to $180°$

$180° - 88° = 92°$

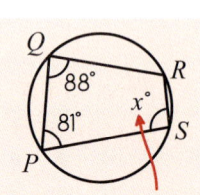

Sometimes we need to combine circle theorems...

e.g. 3 A, B, C and D are points on a circle. PQ is a tangent at the point C.

Angle $PCB = 46°$
Angle $BCD = 102°$

Calculate the size of angle CAD.

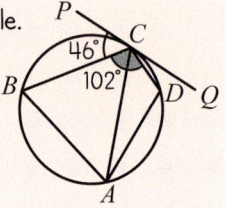

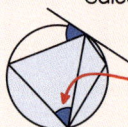

Angle $BAC = 46°$ (alternate segment theorem)

Angle $BAD = 180° - 102° = 78°$ because opposite angles in a cyclic quadrilateral add up to $180°$

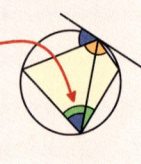

So angle $CAD = 78° - 46° = 32°$

A. Find the value of the labelled angles. Give reasons for your answers.

(i) [circle with A, B, C, D; angles $104°$, $89°$, $a°$]

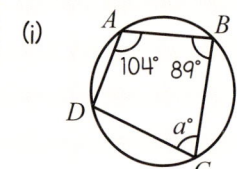

(ii) [circle with D, C, B, A, E; angles $61°$, $93°$, $s°$]

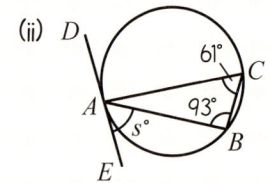

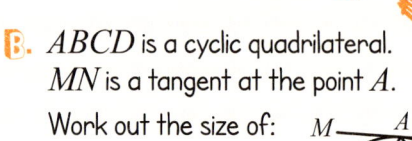

(iii) [circle with A, B, C, D; angles $x°$, $73°$, $y°$, $112°$]

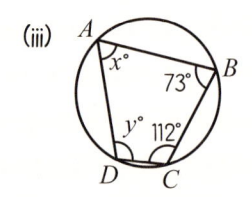

(iv) [circle with Q, R, L, P, M; angles $w°$, $91°$, $v°$, $68°$]

(v) [circle with A, B, C, D, P, Q; angles $79°$, $85°$, $61°$, $b°$]

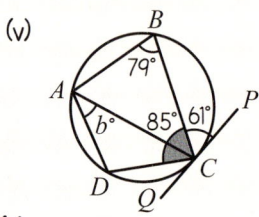

(vi) [circle with M, P, N, S, Q, R; angles $52°$, $h°$, $41°$]

SET?

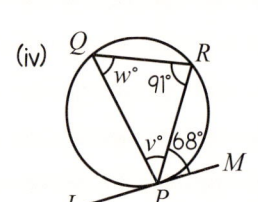

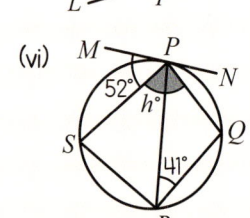

B. $ABCD$ is a cyclic quadrilateral. MN is a tangent at the point A.

Work out the size of:

(i) Angle BAM

(ii) Angle BDC

(iii) Angle BCD

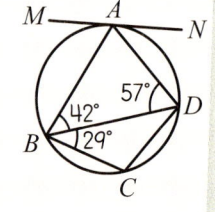

[diagram showing circle with M, A, N, B, C, D; angles $57°$, $42°$, $29°$]

94

Larger diagrams can be found on page 134

GO!

CIRCLE THEOREMS 3

Electric!

1. Alex is solving a circle theorem problem.
   He labels an angle 61° as shown.

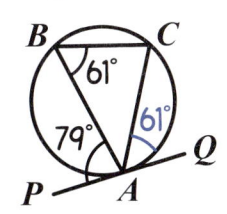

   Alex is not correct.
   Describe the mistake he has made.

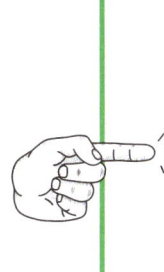

2. A, B, C and D are points on a circle. The line through BD also passes through the centre of the circle. MN is a tangent at A.

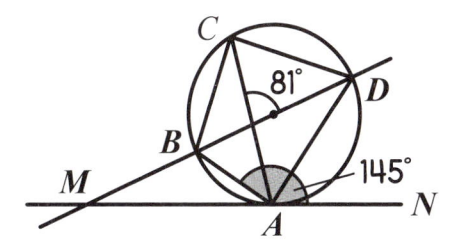

   Work out the size of angle ABC.

3. P, Q, R and S are points on a circle with centre O.

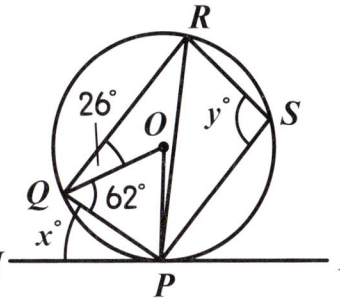

   MN is a tangent at P.

   Explain why $x + y = 120°$

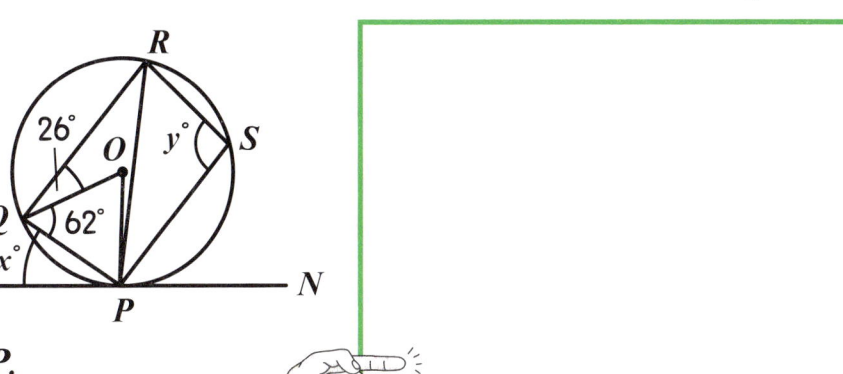

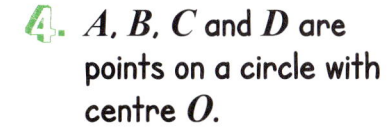

4. A, B, C and D are points on a circle with centre O.

   Prove that

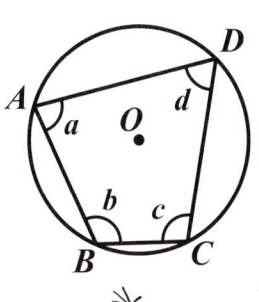

   $a + c = b + d = 180°$

HINT: Draw radii and look for isosceles triangles

READY?

If we think about the equation $x^2 + y^2 = 25$, several integer values for x and y make it true. For example:

$5^2 + 0^2 = 25$, $3^2 + 4^2 = 25$, $(-4)^2 + 3^2 = 25$,
$(-3)^2 + (-4)^2 = 25$, $0^2 + (-5)^2 = 25$, and so on.

Plotting all these points on a grid shows us that they all lie on a circle with its centre at the origin:

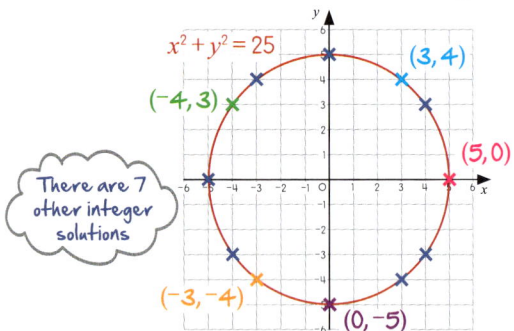

- $x^2 + y^2 = 25$
- $(3, 4)$
- $(-4, 3)$
- $(5, 0)$
- $(-3, -4)$
- $(0, -5)$

There are 7 other integer solutions

All other points on the circle are non-integer solutions to the equation. Every point can be used to create a right-angled triangle with the radius of 5 as its hypotenuse:

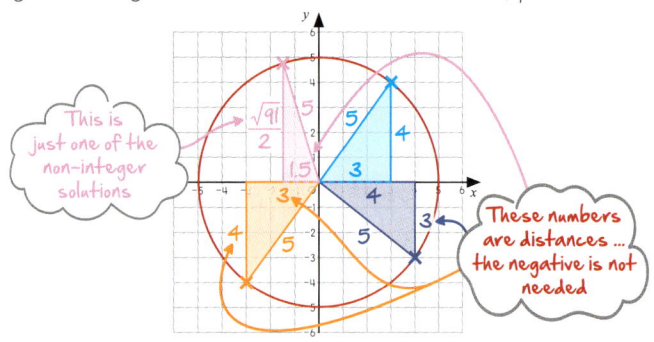

This is just one of the non-integer solutions

These numbers are distances ... the negative is not needed

In general:

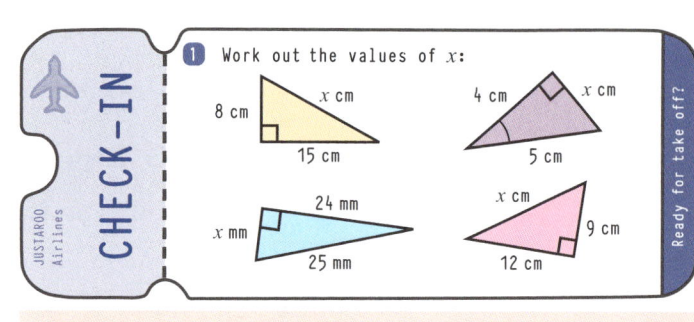

$x^2 + y^2 = r^2$ is a circle with centre $(0, 0)$ and radius r

CHECK-IN

1. Work out the values of x:
   - Triangle: 8 cm, 15 cm, x cm (right angle)
   - Triangle: 4 cm, 5 cm, x cm
   - Triangle: x mm, 24 mm, 25 mm
   - Triangle: x cm, 9 cm, 12 cm

e.g. 1 A circle has equation $x^2 + y^2 = 42.25$
Plot the circle on the grid. Label the points of intersection with the axes.

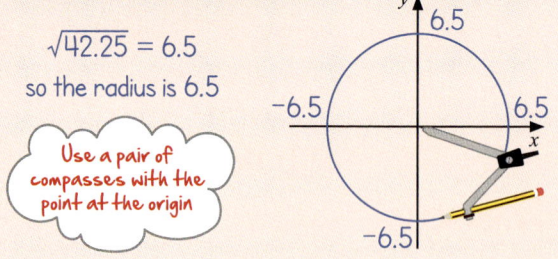

$\sqrt{42.25} = 6.5$
so the radius is 6.5

Use a pair of compasses with the point at the origin

e.g. 2 Here is the graph of a circle.

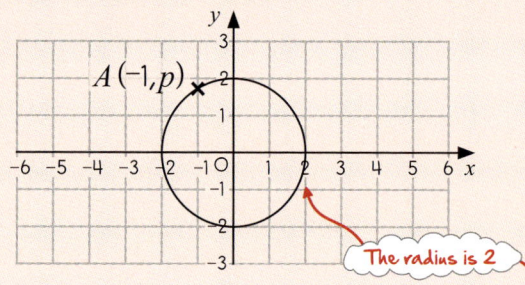

$A(-1, p)$

The radius is 2

The point A lies on the circle and has coordinates $(-1, p)$. Work out the value of p to three significant figures.

The equation of the circle is $x^2 + y^2 = 2^2$
$x = -1$ at point A, so $(-1)^2 + p^2 = 2^2$
Therefore $1 + p^2 = 4$ and $p^2 = 3$
$p = \pm\sqrt{3} = \pm 1.732...$

p must be positive as it is above the x-axis

so $p = 1.73$ to 3 s.f.

SET?

A. State the radius of each circle:
(i) $x^2 + y^2 = 9$
(ii) $x^2 + y^2 = 100$
(iii) $x^2 + y^2 = 12.25$
(iv) $x^2 + y^2 = \dfrac{9}{4}$

B. Write the equation of the circle:

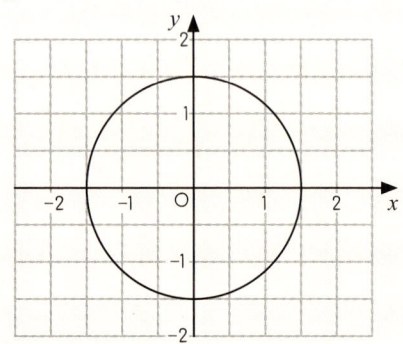

C. Here is the graph of a circle:

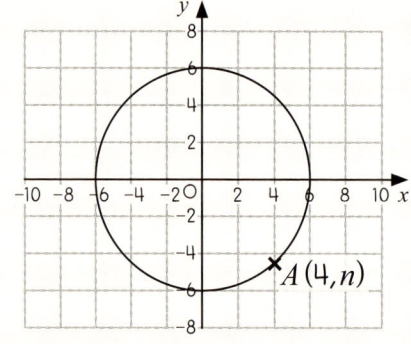

$A(4, n)$

The point A lies on the circle and has coordinates $(4, n)$.
Work out the value of n to two decimal places.

D. Plot the graph of $x^2 + y^2 = 9$ on the grid:

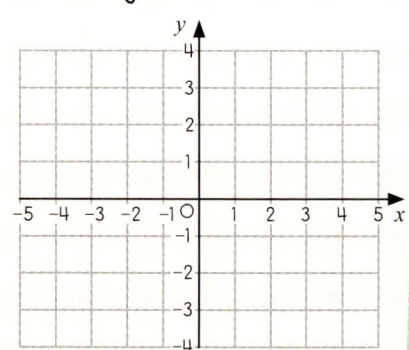

96

GO! EQUATION OF A CIRCLE 1

solid!

1. Jeevan is asked to write the equation of this circle:

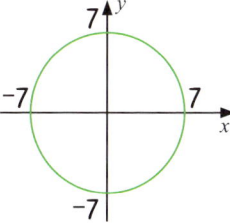

    She writes the answer $x^2 + y^2 = 7$
    Explain why Jeevan is incorrect.

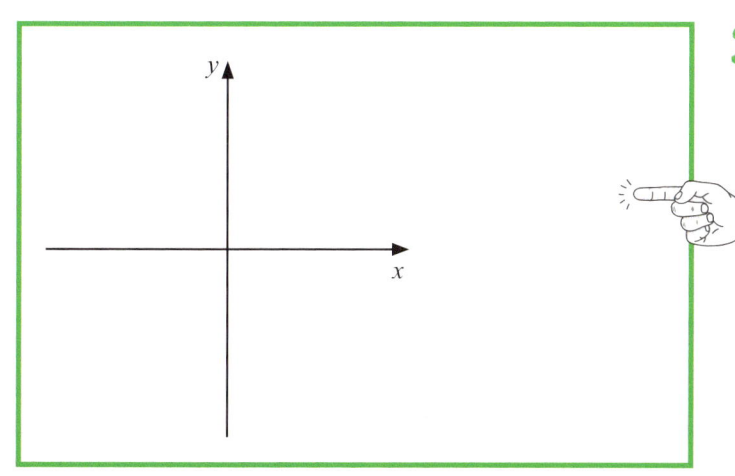

2. a) Find the solutions of the simultaneous equations, giving your solutions to one decimal place:
    $$x^2 + y^2 = 81$$
    $$y = 2x + 5$$

    b) Sketch the graphs of the equations on the grid, labelling the points of intersection with the axes.

3. The point M lies on the circle with equation $x^2 + y^2 = 289$

    a) The coordinates of M are $(a, 15)$ where a is a positive integer. Find the value of a.
    b) N is another point that lies on the circle where both the x-coordinate and y-coordinate are positive integers. Find the coordinates of N.
    c) Find the midpoint of the chord MN.

4. Here are three equations:
    $$x^2 + y^2 = 169$$
    $$2y - x = 29$$
    $$y = x^2 - 13$$

    The graphs of the three equations intersect at a single point.
    Find the coordinates of this point.

READY?

We know that when a radius of a circle meets a tangent, then the radius and tangent make a right angle. This fact can be used to find the equation of a tangent to a circle.

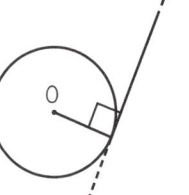

CHECK-IN

1. Find the gradient of the line joining:
   (6,1) and (9,16)   (3,-2) and (5,-6)
   (-7,6) and (-3,7)   (10,3) and (15,-7)

2. Line P has a gradient of $\frac{2}{5}$. Line Q is perpendicular to Line P. Find the gradient of Line Q.

3. A line has a gradient of 5 and passes through the point (2,8). Find the equation of the line.

e.g. 1 The graph shows the circle with equation $x^2 + y^2 = 20$

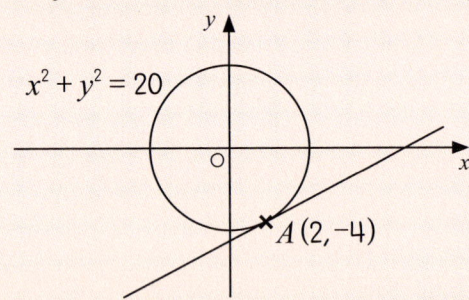

The point A lies on the circle and has coordinates $(2,-4)$. Work out the equation of the tangent to the circle at A.

1 Work out the gradient of the radius OA:
$$\frac{\text{change in } y}{\text{change in } x} = \frac{-4}{2} = -2$$

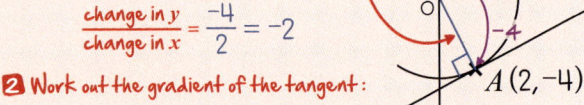

2 Work out the gradient of the tangent:
Gradient of tangent $= -\frac{1}{-2} = \frac{1}{2}$ Negative reciprocal of -2

3 Write $y = mx + c$ for the tangent: $y = \frac{1}{2}x + c$

4 Use the point A to work out the value of c:
At the point $(2,-4)$: $y = \frac{1}{2}x + c \rightarrow -4 = \frac{1}{2} \times 2 + c$
so $-4 = 1 + c$ and $-5 = c$

5 Write the equation of the tangent: $y = \frac{1}{2}x - 5$

e.g. 2 The graph shows the circle with equation $x^2 + y^2 = 58$

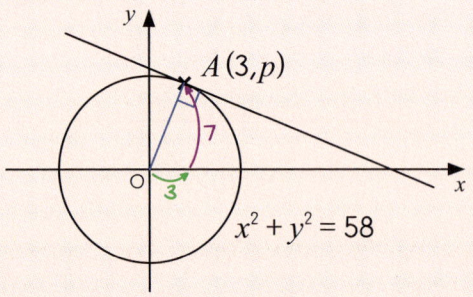

The point A lies on the circle and has coordinates $(3,p)$. Work out the equation of the tangent to the circle at A.

Start by working out the value of p:
$x = 3$ at point A, so $3^2 + p^2 = 58$
Therefore $9 + p^2 = 58$ and $p^2 = 49$
p is positive so $p = 7$

1 Gradient of radius OA: $\frac{\text{change in y}}{\text{change in x}} = \frac{7}{3}$

2 Gradient of tangent $= -\frac{1}{\frac{7}{3}} = -\frac{3}{7}$

3 and 4 At the point $(3,7)$: $y = -\frac{3}{7}x + c \rightarrow 7 = -\frac{3}{7} \times 3 + c$
so $7 = -\frac{9}{7} + c$ and $\frac{58}{7} = c$ **5** Therefore $y = -\frac{3}{7}x + \frac{58}{7}$

SET?

A. The graph shows the circle with equation $x^2 + y^2 = 52$

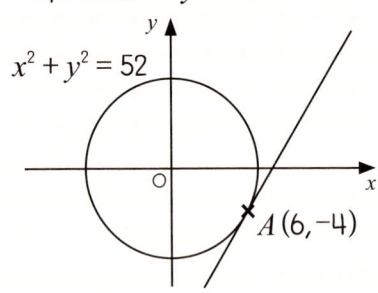

The point A lies on the circle and has coordinates $(6,-4)$.
Work out the equation of the tangent to the circle at A.

B. The graph shows the circle with equation $x^2 + y^2 = 29$

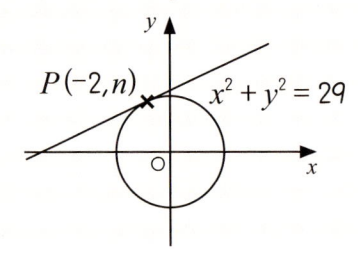

The point P lies on the circle and has coordinates $(-2,n)$.
Work out the equation of the tangent to the circle at P.

C. A circle has equation $x^2 + y^2 = 80$
Find the equation of the two tangents when $x = 4$

GO! EQUATION OF A CIRCLE 2

1. The point A has coordinates $(5,4)$ and lies on a circle with its centre at the origin.

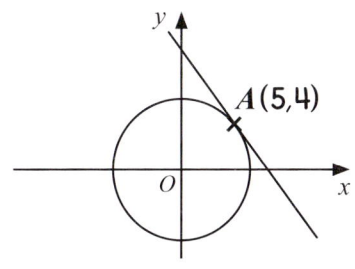

Work out the equation of the tangent at the point A.

2. Here is the circle with equation $x^2 + y^2 = 20$

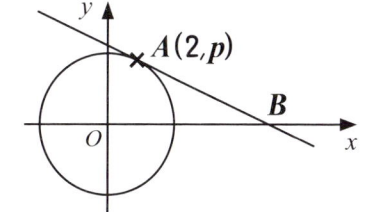

A lies on the circle at the point where $x = 2$. The tangent at A intersects the x-axis at B. Work out the coordinates of B.

3. A circle has equation $x^2 + y^2 = 34$. The centre of the circle is O.

The point P lies on the circle and has coordinates $(5,3)$

The tangent to the circle at P intersects the x-axis at the point Q.

Find the area of triangle OPQ.

4. A circle has its centre at the origin.

A tangent to the circle passes through the points $(2,6)$ and $(20,0)$

Work out the equation of the circle.

READY?

When we enlarge a shape, the scale factor tells us information about how the side lengths change. If we look at the areas of enlarged shapes, a different connection can be seen:

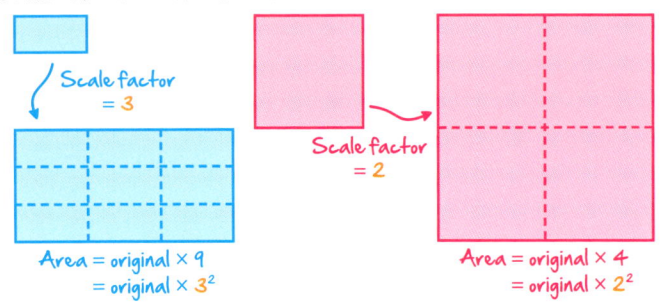

Scale factor = 3
Area = original × 9 = original × 3^2

Scale factor = 2
Area = original × 4 = original × 2^2

In general, if the (length) scale factor is x, then the area scale factor is x^2. There is a different connection between volumes of enlarged shapes:

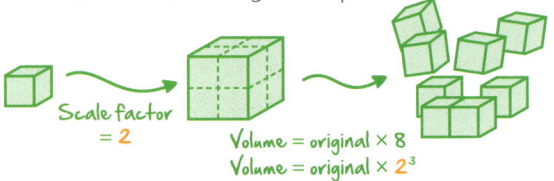

Scale factor = 2
Volume = original × 8
Volume = original × 2^3

In general, if the (length) scale factor is x, then the volume scale factor is x^3. We use these facts to solve problems involving similar shapes in three dimensions.

e.g. 1 Cone A and cone B are similar. Write the ratio of the volume of cone A to the volume of cone B.

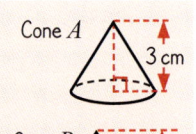

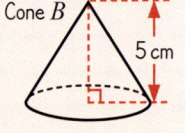

Cone A: 3 cm
Cone B: 5 cm

The ratio of the cone's heights is $3:5$
so the ratio of their volumes is $3^3:5^3$
$= 27:125$

CHECK-IN

1. ABC and PQR are similar triangles.

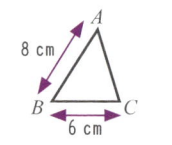

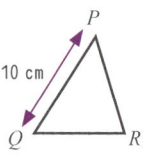

   8 cm, 6 cm, 10 cm

   Find the length of QR.

2. Find the value of m and n if:
   $5:12 = m:36$ $7:10 = n:12$

e.g. 2 Sphere B has a surface area 49 times greater than the surface area of sphere A. How many times greater is the radius of sphere B than the radius of sphere A?

Any two spheres must be mathematically similar

$\sqrt{49} = 7$, so the radius of sphere B is seven times greater

Sometimes more than one step is needed:

e.g. 3 Solid A and solid B are similar. The ratio of the volume of A to the volume of B is $216:125$
Write the ratio of the surface area of A to the surface area of B.

The ratio of the lengths is $\sqrt[3]{216}:\sqrt[3]{125} = 6:5$
so the ratio of the surface areas is $6^2:5^2 = 36:25$

We can solve more complex problems using these ratios:

e.g. 4 The ratio of the surface area of cube P to the surface area of cube Q is $4:81$
The volume of cube P is $24\,cm^3$. Work out the volume of cube Q.

1 Use the fact that two cubes are similar and find the ratio of volumes:
The ratio of the cube's side lengths is $\sqrt{4}:\sqrt{81} = 2:9$
so volume of P : volume of Q is $2^3:9^3 = 8:729$

2 Use the ratio to tackle the problem:
×3 ($8:729$) ×3
   $24:2187$
Volume of cube Q = $2187\,cm^3$

SET?

A. Two pyramids are similar. The ratio of their heights is $3:4$
What is the ratio of their volumes?

B. Two cylinders are similar. The ratio of their base areas is $25:4$
What is the ratio of their heights?

C. Two spheres have volumes in the ratio $125:64$
What is the ratio of their diameters?

D. Two cubes have volumes in the ratio $27:8$
What is the ratio of their surface areas?

E. Two cuboids are mathematically similar.

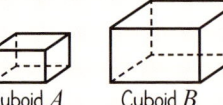

Cuboid A Cuboid B

The ratio of the surface area of cuboid A to the surface area of cuboid B is $9:16$
What is the ratio of the volume of cuboid A to the volume of cuboid B?

F. Two squares have side lengths in the ratio $1:5$. The area of the small square is $10\,cm^2$.
What is the area of the large square?

G. A table tennis ball has a diameter of $40\,mm$.
The diameter of a size 5 netball is 5.5 times greater.
How many times greater is the volume of air in a netball?
Give your answer to the nearest whole number.

GO! SIMILARITY

Incredible!

1. Enamel paint comes in two sizes of cylindrical tin:

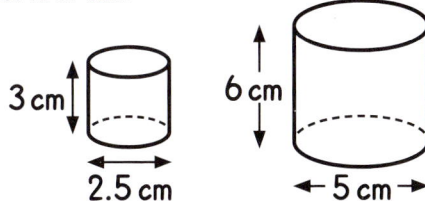

   3 cm, 2.5 cm / 6 cm, 5 cm

   The coverage of the smaller tin is 0.3 m^2. Work out the coverage of the larger tin.

   HINT: The volume of paint in the small tin will cover 0.3m^2

2. The radius of the moon is about one quarter of the radius of the earth.

   a) What fraction of the earth's mass would you expect the moon's mass to be?

   b) Describe an assumption that you have made.

3. Two cones are mathematically similar.
   The base area of cone A is 30 cm^2
   The base area of cone B is 480 cm^2
   Cone B has a volume of 2368 cm^3

   Lucy is asked to work out the volume of cone A. She gives the answer 37 cm^3.

   Do you agree with Lucy? Justify your answer.

4. Solid P is mathematically similar to solid Q. The ratio of the volume of P to the volume of Q is $5:16$

   The surface area of Q is 345 cm^2.

   Calculate the surface area of P correct to three significant figures.

READY?

More complex solids can often be broken down into individual parts to help solve problems.

e.g. 1
A solid is made by joining a cylinder and a cone.

Calculate the surface area of the solid correct to three significant figures.

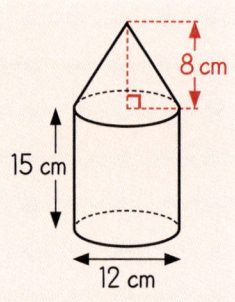

1 Consider each part separately

a The circular base
Radius = 12 ÷ 2 = 6 cm
Area = $\pi \times 6^2$ = 113.097... cm²

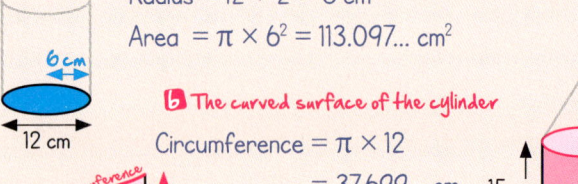

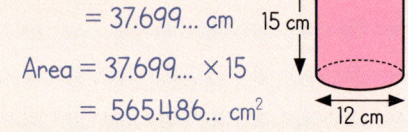

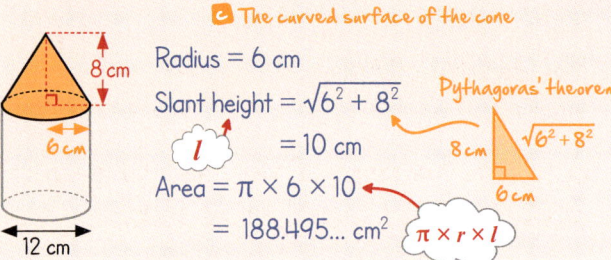

b The curved surface of the cylinder 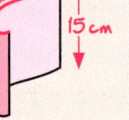 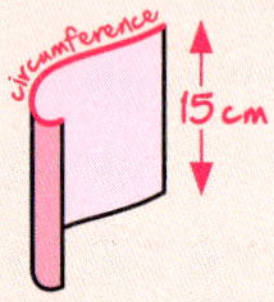
Circumference = $\pi \times 12$
= 37.699... cm
Area = 37.699... × 15
= 565.486... cm²

c The curved surface of the cone
Radius = 6 cm
Slant height = $\sqrt{6^2 + 8^2}$
l = 10 cm
Area = $\pi \times 6 \times 10$
= 188.495... cm²

Pythagoras' theorem: $\sqrt{6^2 + 8^2}$

$\pi \times r \times l$

2 Find the total surface area:
Surface area = 113.097... + 565.486... + 188.495...
= 867.07... ≈ 867 cm² to 3 s.f.

CHECK-IN

1. Find the surface area and volume of the prism: (5 cm, 12 cm, 10 cm)

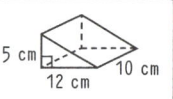

2. Find the surface area of the cone in terms of π: (4 cm, 6 cm)

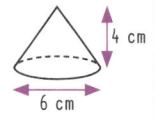

3. Find the surface area of the cylinder to 1 d.p. (5 cm, 8 cm)

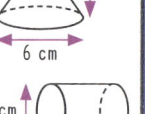

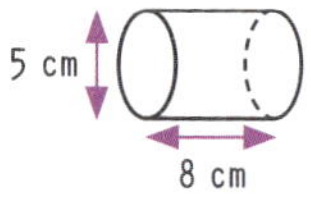

4. Find the volume of the pyramid. (6 cm, 5 cm, 8 cm)

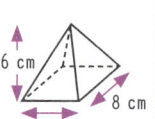

5. Find the exact volume of a sphere with radius 6 cm.

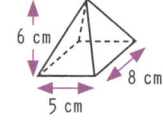

See Key Skills page 6 for formulae support

e.g. 2
A solid is made by joining a cube and half a cylinder. The cube has side length h cm.

Find an expression in terms of π and h for the volume of the solid.

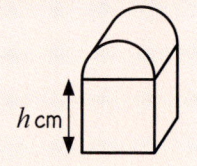

1 Consider each part separately

a The half-cylinder 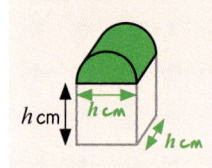
Radius = $h \div 2 = \frac{h}{2}$ cm and length = h
Volume of cylinder = $\pi \times \left(\frac{h}{2}\right)^2 \times h = \frac{1}{4}\pi h^3$
Volume of half-cylinder = $\frac{1}{8}\pi h^3$

b The cube
Volume = h^3

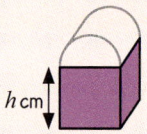

2 Find the total volume:
$\left(\frac{1}{8}\pi h^3 + h^3\right)$ cm³

SET?

A. A solid is made by joining a cylinder and a hemisphere. (6 cm, 5 cm)

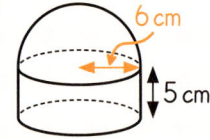

Calculate the volume of the solid in terms of π.

B. A solid is made by joining a cuboid and a pyramid. (9 cm, 4 cm, 6 cm, 6 cm)

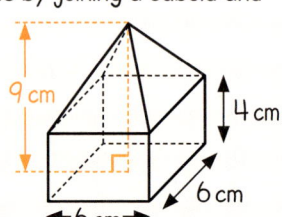

Calculate the volume of the solid.

C. A solid is made by joining a cylinder and a cone. (26 cm, 10 cm, 14 cm)

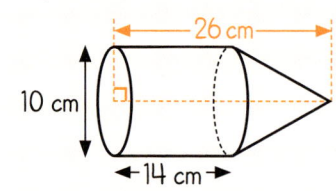

Calculate the surface area of the solid.
Give your answer correct to three significant figures.

D. The cross-section of the solid is a square joined to a quarter-circle: (n cm, $2n$ cm, $2n$ cm)

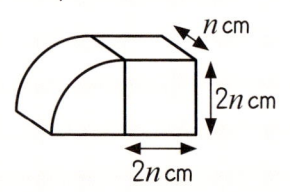

Find an expression for the volume of the solid in terms of n and π.

102

GO! SOLIDS — Good stuff!

1. Cement is stored in a silo. The silo can be modelled as a cylinder joined to a cone as shown.

The silo is 70% full.

Work out the height of the remaining space in the silo.

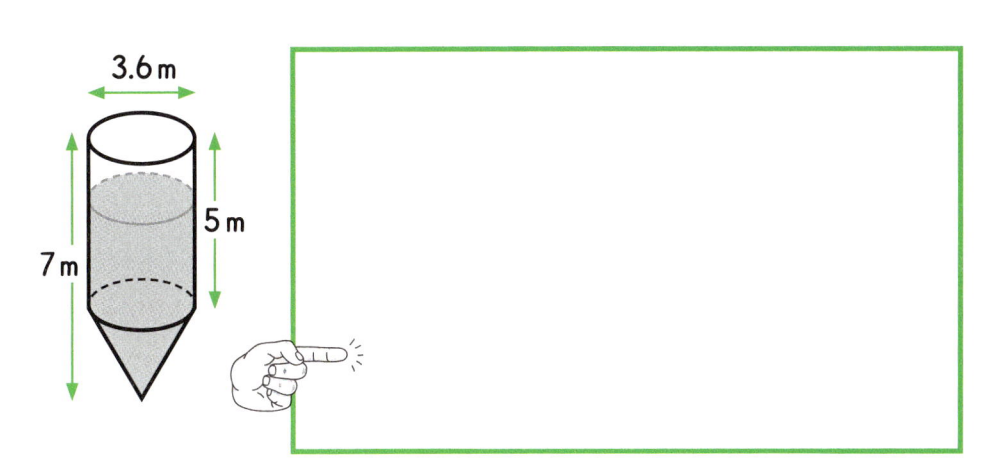

3.6 m
5 m
7 m

2. A solid is made by joining a cylinder and a hemisphere as shown.

12 cm
19 cm

The density of the solid is 1.2 g/cm³.

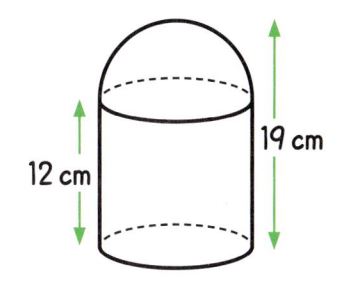

Calculate the mass of the solid. Give your answer to the nearest whole number.

HINT: There is enough information to find the radius of the base

3. A cone is filled with liquid to $\frac{2}{3}$ of its depth.

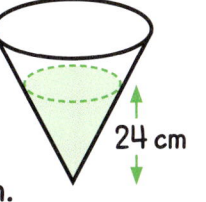

24 cm

The depth of the liquid is 24 cm.

The volume of liquid held in the cone is 900 millilitres to two significant figures.

Work out the radius of the cone to three significant figures.

HINT: Think about similar shapes

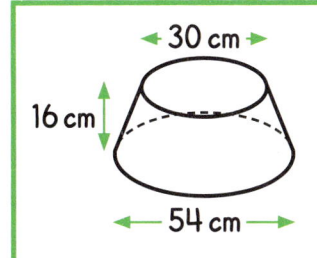

4. A frustum of a cone is the solid that remains when a smaller cone is cut off.

30 cm
16 cm
54 cm

a) Find the exact volume of this frustum.

b) Work out the surface area of the cone that has been cut off.

HINT: Think about similar triangles

READY?

Pythagoras' theorem and trigonometry can be combined to solve problems in 3 dimensions too. It is very helpful to be able to visualise 2D shapes within 3D shapes. For example, we can make different right-angled triangles by joining vertices of a cuboid:

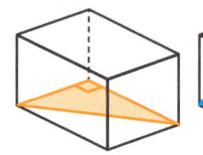

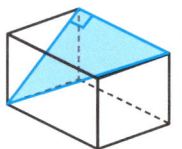

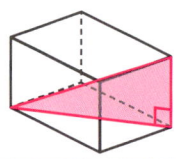

CHECK-IN

1 Work out the values of x to 1 decimal place:

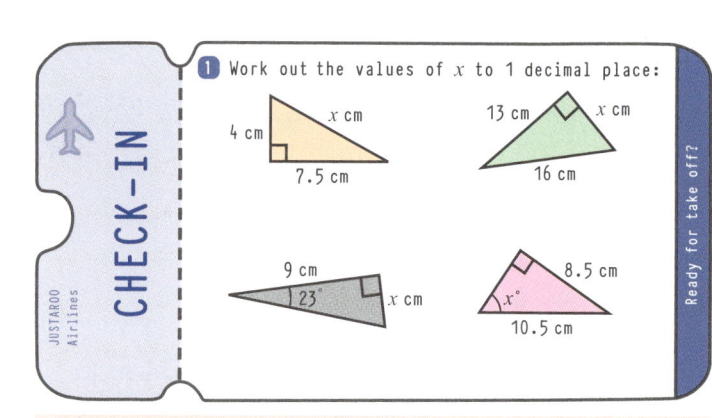

(triangle) 4 cm, 7.5 cm, x cm

(triangle) 13 cm, 16 cm, x cm

(triangle) 9 cm, 23°, x cm

(triangle) 8.5 cm, 10.5 cm, $x°$

---

e.g. 1
$PQRSTU$ is a prism.
The line PU meets the base $PQRS$ at an angle of 22°.

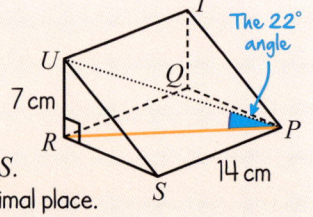

Work out the length of RS. Give your answer to 1 decimal place.

The 22° angle

$UQ = 7$ cm, $SP = 14$ cm

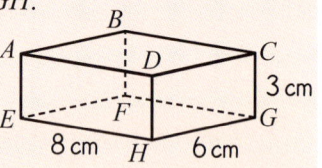

1 Identify the problem: We need to know the length PR to be able to use Pythagoras' theorem with triangle PRS

PR is also a side of triangle PRU

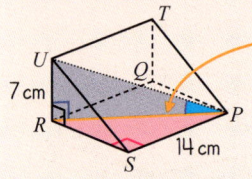

2 Extract triangle PRU and work out length PR:

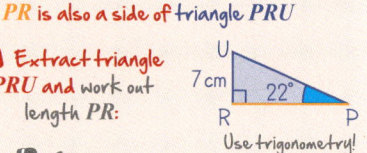

Use trigonometry!

a Label: $O = 7$ cm, 22°, A (triangle RPU)

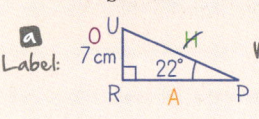

b Choose: We know O and need A, so use $\tan\theta = \dfrac{O}{A}$

c Substitute: $O = 7$ cm and $A = PR$ → $\tan 22° = \dfrac{7}{PR}$

d Solve: $PR = \dfrac{7}{\tan 22°} = 17.325...$

3 Extract triangle PRS and work out length RS:

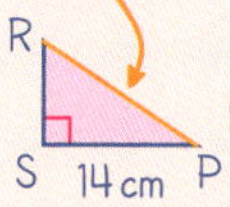

$17.325...^2 = RS^2 + 14^2$ so $\sqrt{17.325...^2 - 14^2} = RS$

so $RS = 10.206... = 10.2$ cm to 1 d.p.

Do NOT round until the end of the problem

---

e.g. 2
$ABCDEFGH$ is a cuboid. Calculate the angle that CE makes with the plane $EFGH$.

(cuboid with $EH = 8$ cm, $HG = 6$ cm, $CG = 3$ cm)

1 Identify the problem:

The required angle (θ)

Plane $EFGH$ (base of the cuboid)

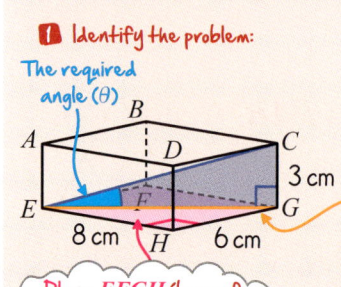

We need to find the length EG to be able to use trigonometry with triangle CEG

EG is also the hypotenuse of triangle EGH

2 Extract triangle EGH and work out length EG:

Use Pythagoras' theorem!

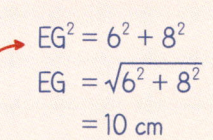

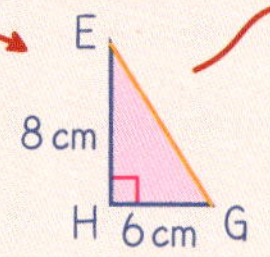

$EG^2 = 6^2 + 8^2$
$EG = \sqrt{6^2 + 8^2}$
$= 10$ cm

3 Extract triangle CEG and work out angle CEG:

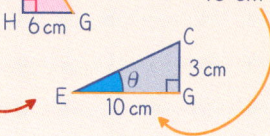

a Label: H, θ, $O = 3$ cm, A

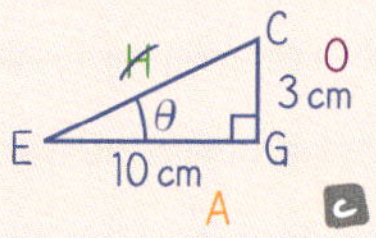

b Choose: We know O and A so use $\tan\theta = \dfrac{O}{A}$

c Substitute: $\tan\theta = \dfrac{3}{10}$

d Solve: $\theta = \tan^{-1}\left(\dfrac{3}{10}\right) = 16.699...°$
$\theta = 16.7°$ (to 1 d.p.)

SET?

A. $ABCDEFGH$ is a cuboid.

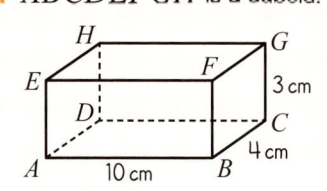

(cuboid with $AB = 10$ cm, $BC = 4$ cm, $CG = 3$ cm)

Work out:
(i) The length AF

(ii) Angle BDH

(iii) The length BH

B. $PQRSTU$ is a prism.

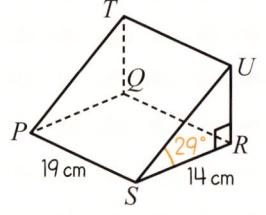

(prism with $PS = 19$ cm, $SR = 14$ cm, angle 29°)

Work out:
(i) The length QT

(ii) The length PR

(iii) Angle RPU

C. Here is a pyramid with a square base $ABCD$. V is directly above the midpoint of AC.

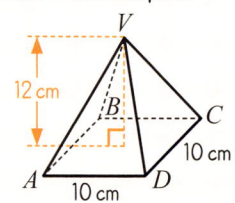

(pyramid, height 12 cm, $AD = 10$ cm, $DC = 10$ cm)

(i) Work out the length AC

(ii) Find the size of angle CAV

Give answers to one decimal place if needed

104

GO! TRIGONOMETRY IN 3D (AND PYTHAGORAS' THEOREM)

1. AB is the diameter of a circular base of a cone. The vertex P is directly above the centre of the base.

    $AB = 9$ cm
    $AP = 11$ cm

    Find the size of angle PAB.

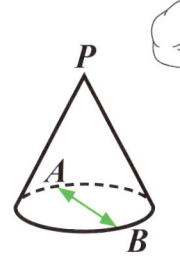

    Give your answer to three significant figures

2. Here is a cuboid:

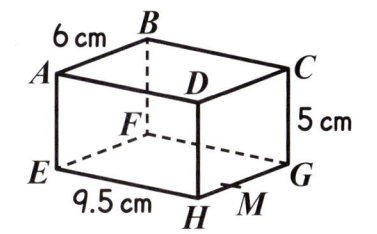

    M is the point such that $GM:MH = 2:1$

    Calculate the size of angle AME.
    Give your answer to one decimal place.

3. The point X is at the centre of the square base of a pyramid $ABCDE$.

    M is the midpoint of CD.

    Angle $EMX = 66°$

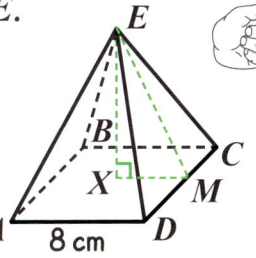

    Calculate the surface area of the pyramid.

    Give your answer to the nearest cm²

4. Here is a prism:

    Angle $CGH = 68°$
    $AD = 6$ cm
    $CG = 8$ cm
    $DH = 3$ cm

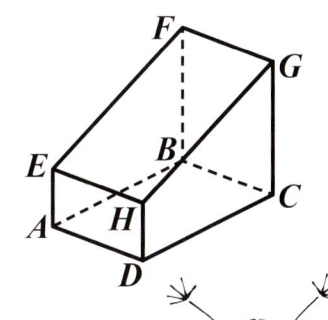

    Calculate the size of angle ECA to one decimal place.

READY?

We often label non-right angled triangles like this ...

Capital letters for corners ... ...matched with the lower case letter on the opposite side

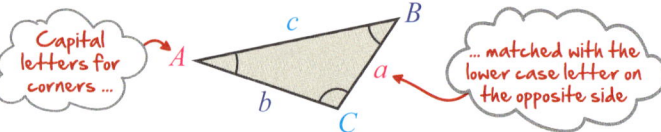

The capital letters are used for the angles too

By splitting a non-right triangle into two right-angled triangles we can find other facts:

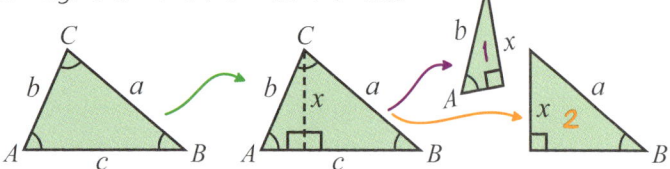

In triangle 1: $\sin A = \frac{x}{b}$ so $b \sin A = x$

In triangle 2: $\sin B = \frac{x}{a}$ so $a \sin B = x$

So: $b \sin A = a \sin B$ and also: $\frac{b}{\sin B} = \frac{a}{\sin A}$

We can also show that: $\frac{a}{\sin A} = \frac{b}{\sin B} = \frac{c}{\sin C}$ ← *The sine rule*

The **sine rule** can be used to find missing sides ...

e.g. 1 Find the value of x in this triangle.

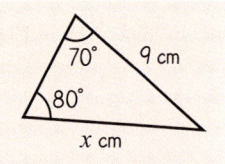

1. Label the sides and angles:
2. Look for useful pairs and choose part of the sine rule: $\frac{a}{\sin A} = \frac{b}{\sin B} = \frac{c}{\sin C}$ 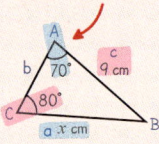
3. Substitute: $\frac{x}{\sin 70°} = \frac{9}{\sin 80°}$ so $x = \frac{9}{\sin 80°} \times \sin 70°$
4. Solve: $= 8.59$ (3 s.f.) *sensible accuracy*

CHECK-IN

1. Work out the values of x to 1 decimal place:

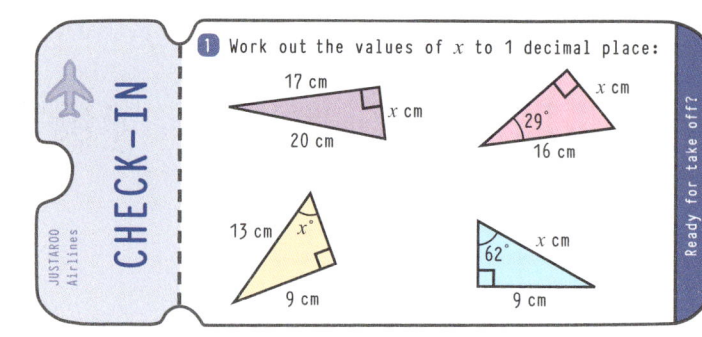

We can also find missing angles using the sine rule ...

e.g. 2 Find the value of y in this triangle.

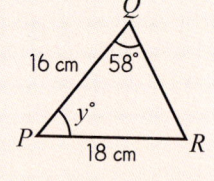

1. Label: 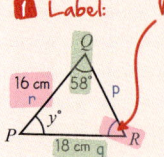 *We need to find this angle first*
2. Pair and choose: $\frac{p}{\sin P} = \frac{q}{\sin Q} = \frac{r}{\sin R}$
3. Substitute: $\frac{18}{\sin 58°} = \frac{16}{\sin R}$ so $21.225... = \frac{16}{\sin R}$
4. Rearrange and solve: $21.225... \times \sin R = 16$ so $\sin R = \frac{16}{21.225...}$
   and $R = \sin^{-1}\left(\frac{16}{21.225...}\right) = 48.9°$ (3 s.f.)

Therefore $y = 180 - 58 - 48.9 = 73.1$ (3 s.f.)

Sometimes we need to use the sine rule more than once:

e.g. 3 $ABCD$ is a quadrilateral. Find the size of angle BAC.

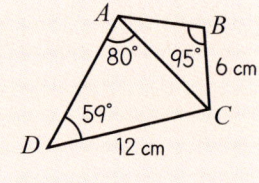

Work with triangle ACD first ...

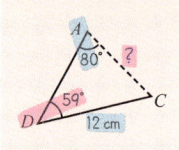

$\frac{AC}{\sin 59°} = \frac{12}{\sin 80°}$

so $AC = \frac{12}{\sin 80°} \times \sin 59° = 10.44...$

... and then with triangle ABC: $\frac{10.44...}{\sin 95°} = \frac{6}{\sin A}$

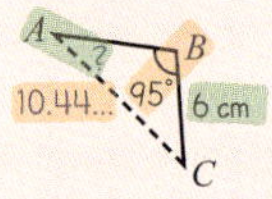

so $\sin A = \frac{6}{10.48...}$ $10.44... \div \sin 95° = 10.48...$

angle BAC → $A = \sin^{-1}\left(\frac{6}{10.48...}\right) = 34.9°$ (3 s.f.)

SET?

A. Find the value of the sides labelled with letters. Give answers to 3 significant figures.

(i) (ii) 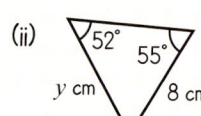 (iii)

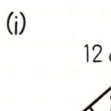

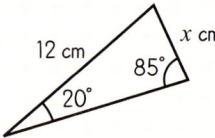

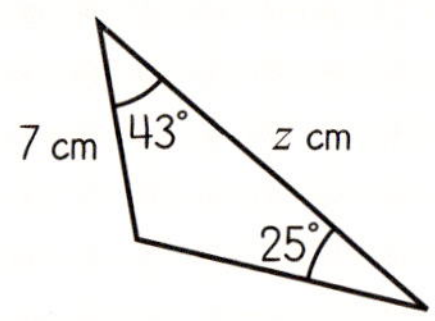

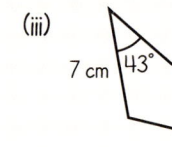

B. Find the value of the angles labelled with letters. Give answers to one decimal place.

(i) (ii) 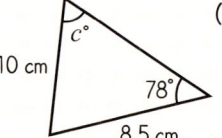 (iii)

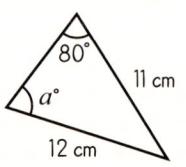

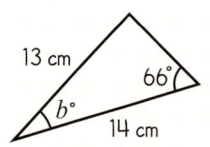

C. Work out the value of angle PSQ to one decimal place.

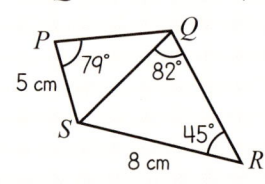

106

GO! THE SINE RULE

Love it!

1. Hattie is given this triangle:

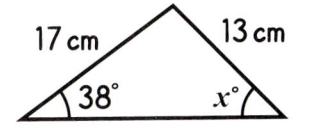

   (Triangle with sides 17 cm, 13 cm, angle 38°, and angle x°)

   She writes:

   $$\frac{\sin x°}{17} = \frac{\sin 38°}{13}$$

   Do you agree with Hattie? Explain why.

2. In triangle PQR:
   $QR = 12.5$ cm
   $PR = 16$ cm
   Angle $QPR = 32°$
   Find the size of angle PQR.

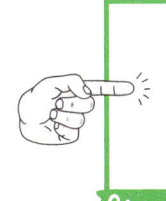

   Give your answer to three significant figures

3. Find the value of x.

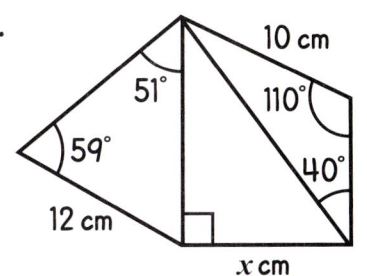

   (Triangle diagram with angles 51°, 110°, 59°, 40°, sides 10 cm, 12 cm, and x cm)

   Give your answer to three decimal places

4. The diagram shows the position of three towns on a map.

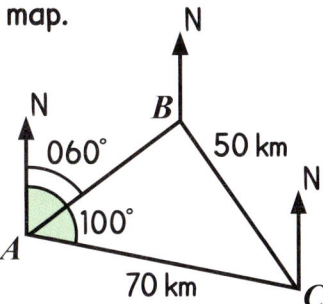

   (Diagram showing towns A, B, C with bearing 060° at A, angle 100° at A, 50 km from B, 70 km from A to C)

   Find the bearing of C from B.

   Give your answer to the nearest degree

5. Here is an obtuse-angled triangle.

   (Triangle LMN with NL = 14.2 cm, LM = 9.5 cm, angle at N = 29°)

   Calculate the size of angle LMN to one decimal place.

   HINT: Notice that angle LMN is obtuse

READY?

We have already seen that we can label sides and angles in non-right angled triangles like this.

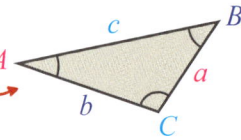

It is possible to show that:

$$a^2 = b^2 + c^2 - 2bc\cos A$$

This is called the **cosine rule**. It is also true that:

Notice the pattern!

$$b^2 = a^2 + c^2 - 2ac\cos B$$
$$c^2 = a^2 + b^2 - 2ab\cos C$$

The pattern also helps us write the cosine rule when the triangle is labelled with different letters: e.g.

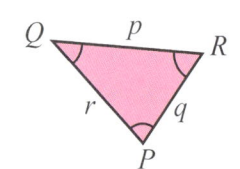

$$p^2 = q^2 + r^2 - 2qr\cos P$$

We often use the cosine rule if the problem involves three sides and one angle (or when the sine rule does not work!). The angle is the starting point...

CHECK-IN

1. Work out the values of x to 1 decimal place:
   - Triangle with 17 cm, 23°, 73°, x cm
   - Triangle with 84 mm, 150 mm, 34°, $x°$

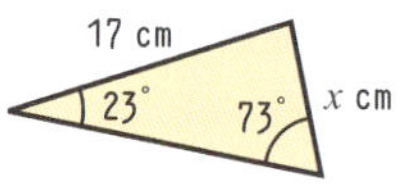

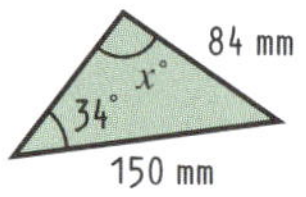

2. Sketch and label triangle ABC with:
   - $AB = 7.6$ cm
   - $BC = 8.3$ cm
   - Angle $ACB = 58°$

e.g. 1 Find the value of x in this triangle.

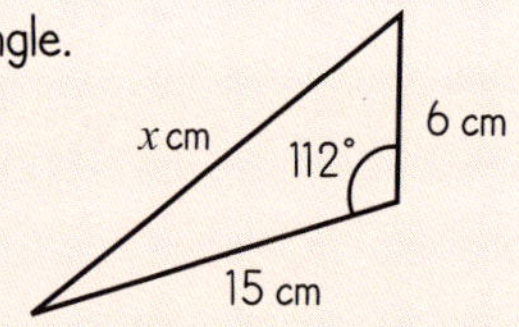

1. Label the sides and angles.

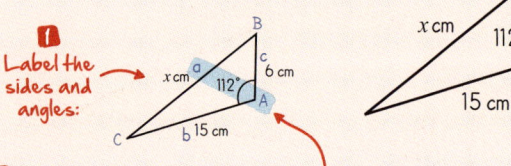

2. Look for the angle between two sides; write down the cosine rule:
   $$a^2 = b^2 + c^2 - 2bc\cos A$$

3. Substitute: $x^2 = 15^2 + 6^2 - 2 \times 15 \times 6 \times \cos 112°$

4. Solve: $x^2 = 225 + 36 - 180 \times -0.374...$
   $x^2 = 328.429...$
   $x = 18.122... = 18.1$ (3 s.f.)

Lengths must be the positive square root

e.g. 2 Find the exact value of b in this triangle.

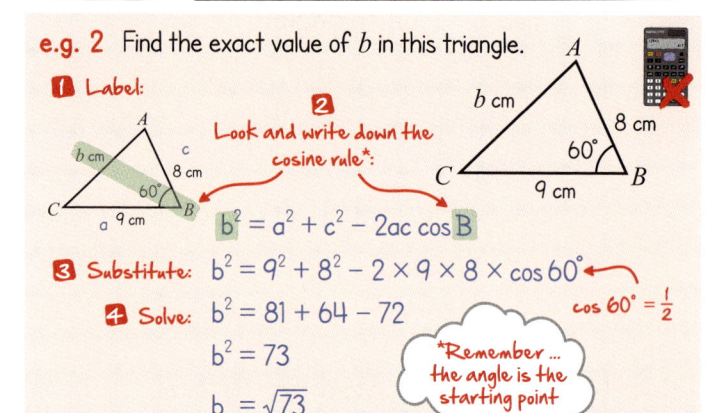

1. Label.

2. Look and write down the cosine rule*:
   $$b^2 = a^2 + c^2 - 2ac\cos B$$

3. Substitute: $b^2 = 9^2 + 8^2 - 2 \times 9 \times 8 \times \cos 60°$

4. Solve: $b^2 = 81 + 64 - 72$
   $b^2 = 73$
   $b = \sqrt{73}$

$\cos 60° = \frac{1}{2}$

Remember... the angle is the starting point

We can also use the cosine rule to find missing angles...

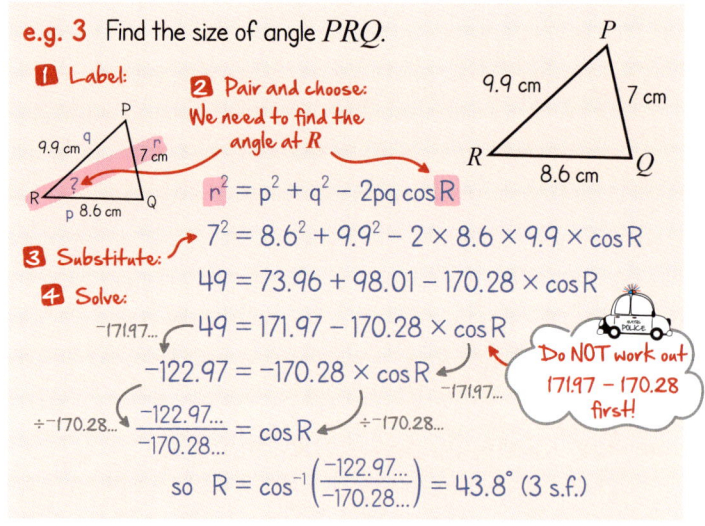

e.g. 3 Find the size of angle PRQ.

1. Label.

2. Pair and choose: We need to find the angle at R
   $$r^2 = p^2 + q^2 - 2pq\cos R$$

3. Substitute: $7^2 = 8.6^2 + 9.9^2 - 2 \times 8.6 \times 9.9 \times \cos R$

4. Solve: $49 = 73.96 + 98.01 - 170.28 \times \cos R$
   $49 = 171.97 - 170.28 \times \cos R$
   $-122.97 = -170.28 \times \cos R$
   $\dfrac{-122.97...}{-170.28...} = \cos R$
   so $R = \cos^{-1}\left(\dfrac{-122.97...}{-170.28...}\right) = 43.8°$ (3 s.f.)

Do NOT work out $171.97 - 170.28$ first!

SET?

A. Find the value of the labelled sides. Give answers to 3 significant figures.

(i) Triangle: 11 cm, 70°, 13 cm, p cm

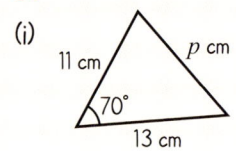

(ii) Triangle: 12 cm, 50°, 7 cm, q cm

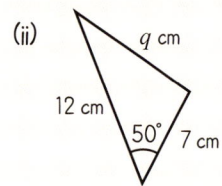

(iii) Triangle: 8 cm, 115°, 14 cm, r cm

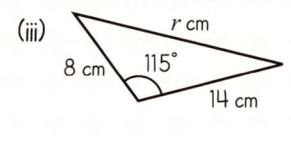

B. Find the value of the labelled angles. Give answers to one decimal place.

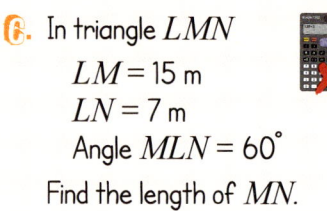

(i) Triangle: 14 cm, 23 cm, 25 cm, $x°$

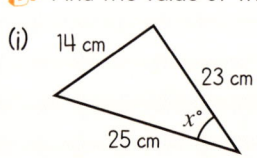

(ii) Triangle: 18 cm, 16 cm, 13 cm, $y°$

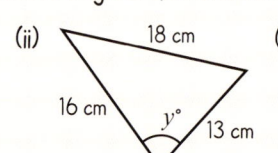

(iii) Triangle: 20 m, 12 m, 17 m, $z°$

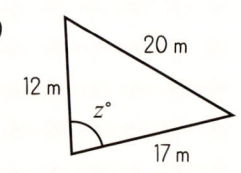

C. In triangle LMN
- $LM = 15$ m
- $LN = 7$ m
- Angle $MLN = 60°$

Find the length of MN.

108

THE COSINE RULE

GO!

Supreme!

1. Fintan is given this triangle.

   13 cm, x°, 6 cm, 9 cm

   He writes:

   $13^2 = 6^2 + 9^2 - 2 \times 6 \times 9 \times \cos x$
   $169 = 36 + 81 - 108 \times \cos x$
   $169 = 9 \times \cos x$
   $x = \cos^{-1}\left(\frac{169}{9}\right)$

   Fintan is wrong. Explain his mistake.

2. Find the value of x.

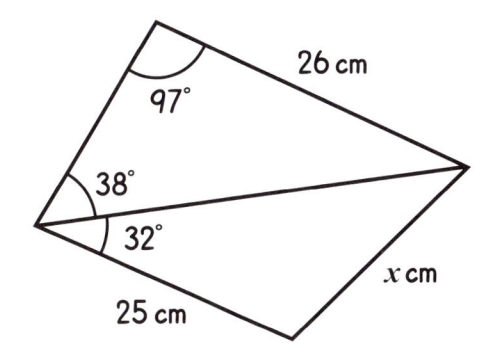

   26 cm, 97°, 38°, 32°, 25 cm, x cm

   Give your answer to three significant figures

3. Find all the angles in this triangle:

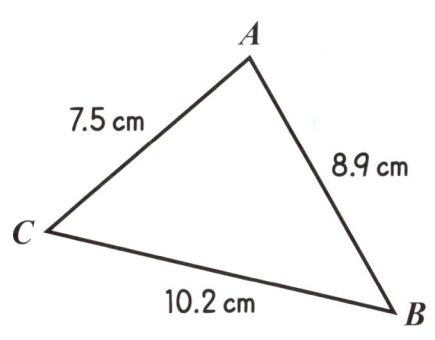

   A, 7.5 cm, 8.9 cm, C, 10.2 cm, B

   Give your answers to one decimal place

4. The diagram shows the distances between three towns.

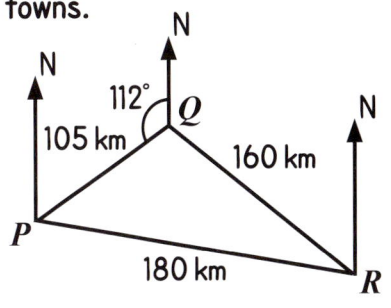

   112° at Q, 105 km, 160 km, P, 180 km, R

   Find the bearing of R from P.

   Give your answer to the nearest degree

READY?

If we don't know the base or perpendicular height of a triangle, there may be other ways to calculate its area. For example, we might know some of the sides or angles:

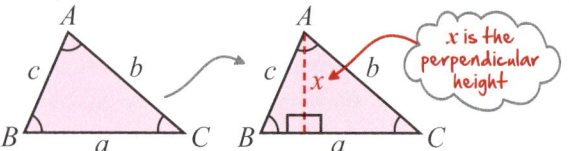

x is the perpendicular height

We can use trigonometry to find the value of x ...

$$\sin C = \frac{\text{opposite}}{\text{hypotenuse}} \rightarrow \sin C = \frac{x}{b} \rightarrow b \sin C = x$$

Remember:
Area of triangle $= \frac{1}{2} \times$ base $\times$ perpendicular height

So:
Area of triangle $= \frac{1}{2} \times a \times b \sin C \rightarrow$ Area $= \frac{1}{2} ab \sin C$

We can use this formula if the problem involves any two sides and the angle between them.

CHECK-IN

1. Work out the values of x to 1 decimal place:
   (triangle: 5.3 cm, 85°, x°, 9.4 cm)
   (triangle: 29°, 82°, x cm, 21 cm)

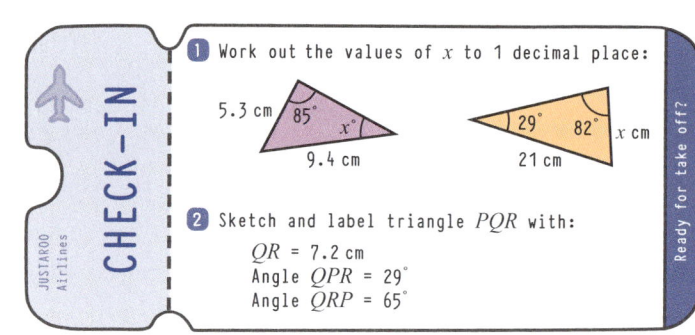

2. Sketch and label triangle PQR with:
   $QR = 7.2$ cm
   Angle $QPR = 29°$
   Angle $QRP = 65°$

We can also find missing sides and angles in triangles ...

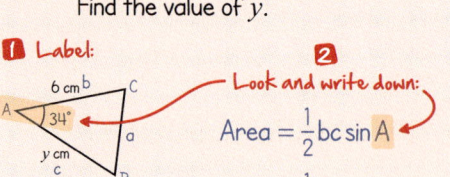

e.g. 1 Find the area of triangle ABC.
Give your answer to one decimal place.

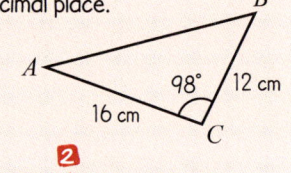

(triangle ABC: angle 98° at A, AB = 12 cm, AC = 16 cm)

1. **Label the sides:** (a = 12 cm, b = 16 cm, C = 98°)

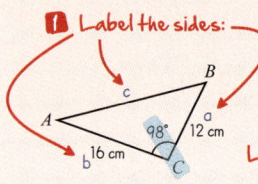

2. Look for the angle between two sides and write down the formula:
   $$\text{Area} = \frac{1}{2} ab \sin C$$

3. **Substitute:** $= \frac{1}{2} \times 12 \times 16 \times \sin 98°$

4. **Work out:** $= 95.06... = 95.1$ cm² (1 d.p.)

e.g. 2 The triangle has an area of 13 cm². Find the value of y.
(triangle: 6 cm, 34°, y cm)

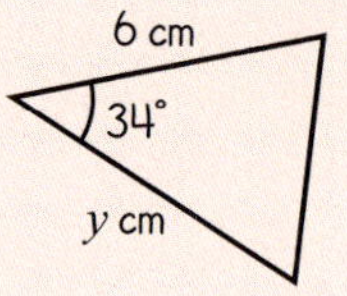

1. **Label:** (b = 6 cm, A = 34°, c = y cm)

2. Look and write down:
   $$\text{Area} = \frac{1}{2} bc \sin A$$

3. **Substitute:** $13 = \frac{1}{2} \times 6 \times y \times \sin 34°$

4. **Solve:** $13 = 1.677... \times y$
   $\div 1.677... \quad 7.749... = y \quad \div 1.677...$

   ($\frac{1}{2} \times 6 \times \sin 34° = 1.677...$)

   So $y = 7.75$ cm (3 s.f.) *sensible accuracy*

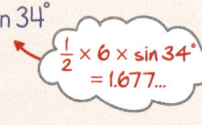

e.g. 3 The triangle has an area of 279 cm². Find the size of angle PQR.
(triangle PQR: PR = 32 cm, RQ = 36 cm)

1. (p = 36 cm, q = 32 cm, r = ?)

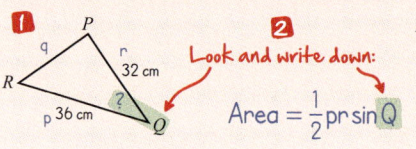

2. Look and write down:
   $$\text{Area} = \frac{1}{2} pr \sin Q$$

3. **Substitute:** $279 = \frac{1}{2} \times 36 \times 32 \times \sin Q$

4. **Solve:** $279 = 576 \times \sin Q$
   $\div 576 \quad \frac{279}{576} = \frac{31}{64} \quad \frac{31}{64} = \sin Q \quad \div 576$

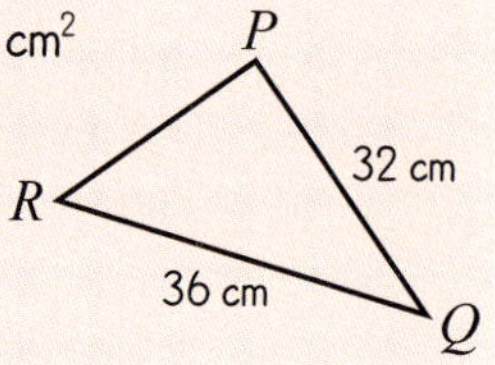

   So angle $PQR = \sin^{-1}\left(\frac{31}{64}\right) = 29.0°$ (3 s.f.)

SET?

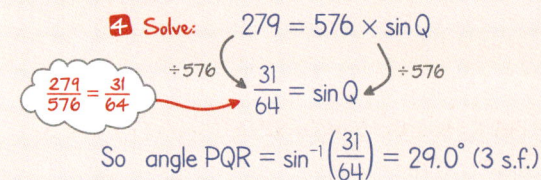

A. Find the area of the triangles. Give answers to 3 significant figures.

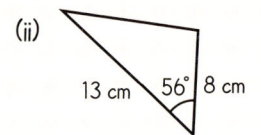

(i) triangle: 19 m, 75°, 14 m
(ii) triangle: 13 cm, 56°, 8 cm
(iii) triangle: 30 cm, 102°, 22 cm

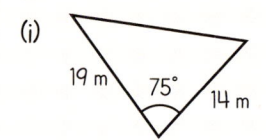

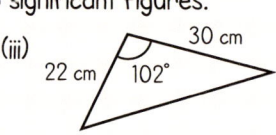

B. Find the value of x. Give answers to one decimal place.

(i) triangle: 12 cm, 64°, x cm, Area = 72 cm²
(ii) triangle: 1.2 m, $x°$, 90 cm, Area = 0.537 m²
(iii) triangle: 107°, x cm, x cm (isosceles), Area = 16 cm²

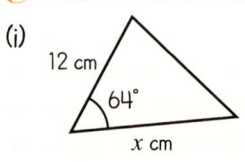

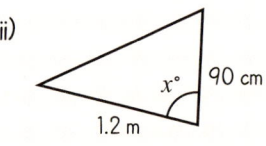

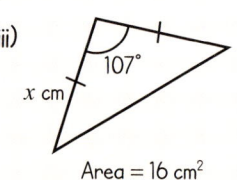

C. In triangle XYZ
$XZ = 16$ cm
$YZ = 7$ cm
Angle $XZY = 30°$

Find the area of the triangle.

110

GO!

AREA OF A TRIANGLE

Fantastic!

1. In triangle ABC

    $AB = 30$ m
    $BC = 24$ m

    The area of the triangle is $180\sqrt{3}$

    Find the size of the acute angle ABC.

2. Find the area of triangle LMN to one decimal place.

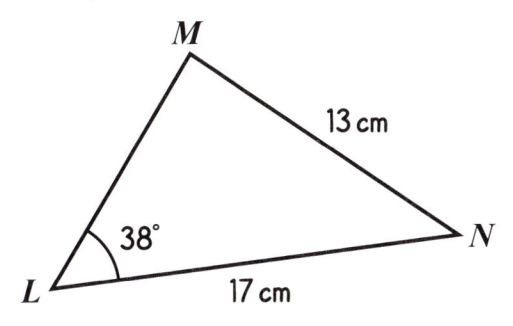

    M, 13 cm, 38°, N, 17 cm, L

    HINT: Use the sine rule first

3. Find the area of quadrilateral $ABCD$.

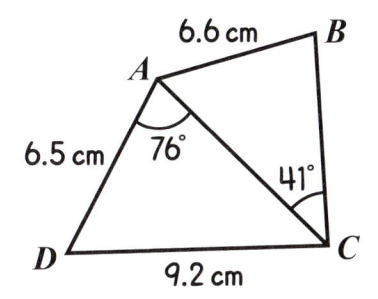

    6.6 cm, A, B, 6.5 cm, 76°, 41°, D, 9.2 cm, C

    Give your answer to two significant figures.

4. Here is a right-angled triangled:

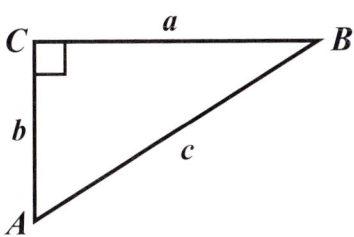

    C, a, B, b, c, A

    Is it true that: Area $= \frac{1}{2}ab\sin C$?

    Explain why.

READY?

We can select and use geometrical knowledge and skills to construct geometric proofs. These often involve the conditions for **congruent** triangles...

CHECK-IN

1. Tick the conditions for congruent triangles:
   SSS   SSA   ASA   SAS   RHS   AAA

2. Work out the size of the interior angle in a regular octagon.

3. ABC and DBE are similar triangles. Find the length of AD.
   (6 cm, 8 cm, 10 cm)

e.g. 1

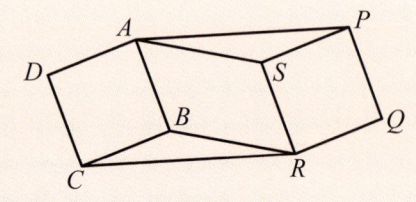

$ABCD$ and $PQRS$ are squares.
$ASRB$ is a parallelogram.
Prove that triangles APS and RCB are congruent.

1. Identify the relevant triangles:

2. State any pairs of equal sides, with reasoning:

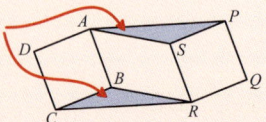

$AS = BR$ as opposite sides of a parallelogram are equal.

Also, the squares $ABCD$ and $PQRS$ must have equal side lengths as $ASRB$ is a parallelogram.

So $BC = PS$.

Do NOT assume that $AP = CR$

3. Find any pairs of equal angles, with reasoning:

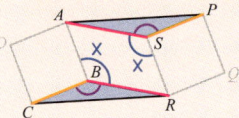

$\angle ASR = \angle RBA$ *'Angle ASR'* as opposite sides of a parallelogram are equal. Label these angles x.

$\angle PSR = \angle ABC = 90°$ since $ABCD$ and $PQRS$ are squares.

So $\angle ASP = \angle RBC$ $(= 360° - 90° - x)$

4. Make a conclusion using conditions for congruent triangles:

Therefore triangles APS and RCB are congruent by **SAS**

e.g. 2

$LMNP$ is an isosceles trapezium.
Angle PLM = Angle LPN
$LM = PN$
Prove that $LN = PM$.

Consider triangles LMN and MNP:
It is given that $LM = PN$.
MN is a shared side.
Co-interior angles add to 180° and $\angle PLM = \angle LPN$, so $\angle LMN = \angle PNM$.
Therefore triangles LMN and MNP are congruent by **SAS**.
Therefore $LN = PM$

We can prove that triangles are **similar** by checking that all three angles are the same.

e.g. 3

ABC is a triangle.
A, D and C lie on a straight line.
Prove that triangles ABC and BCD are similar.

In triangle BCD, $\angle CBD = 180° - 90° - 58° = 32°$

In triangle ABC, $\angle CBA = 180° - 90° - 32° = 58°$

The angles in triangles ABC and BCD are the same (32°, 58° and 90°), so the triangles are similar.

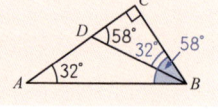

SET?

A.

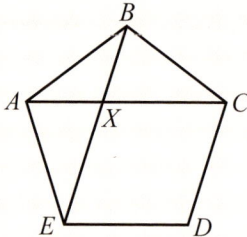

$ABCDE$ is a regular pentagon.
The lines AC and BE intersect at X.
Prove that triangles BCX and AEX are congruent.

B.

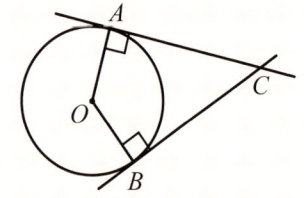

AC and BC are tangents to a circle with centre O.
The tangents intersect at C.
Use congruent triangles to prove that $AC = BC$.

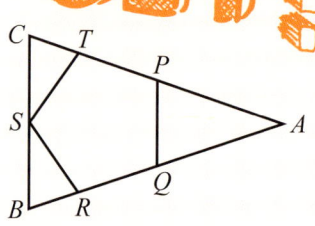

C.

ABC is an isosceles triangle.
A regular pentagon is made by joining five points on ABC.
Prove that triangles PAQ and CST are similar.

GO!

GEOMETRIC PROOF

Unreal!

1. ABC is a triangle with $AB = AC$.
M is a point on AB and N is a point on AC.
Angle $ANB = 102°$
Angle $BMC = 78°$

Prove that triangles ABN and ACM are congruent.

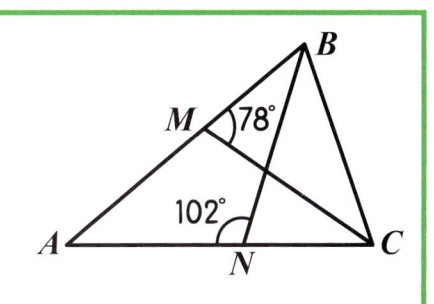

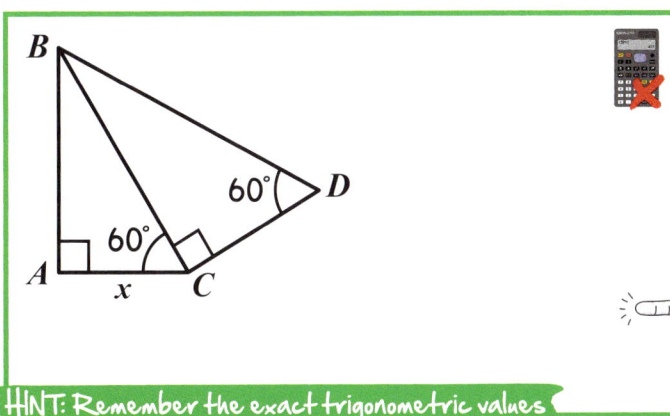

2. Two similar triangles, ABC and BCD, are joined to make a quadrilateral.

Show that the area of quadrilateral $ABDC$ is given by:
$$\frac{7\sqrt{3}}{6}x^2$$

HINT: Remember the exact trigonometric values

3. $ABCD$ is a quadrilateral.

E is the point on BC such that triangle ADE is isosceles.

Prove that triangles ABE and ECD are congruent.

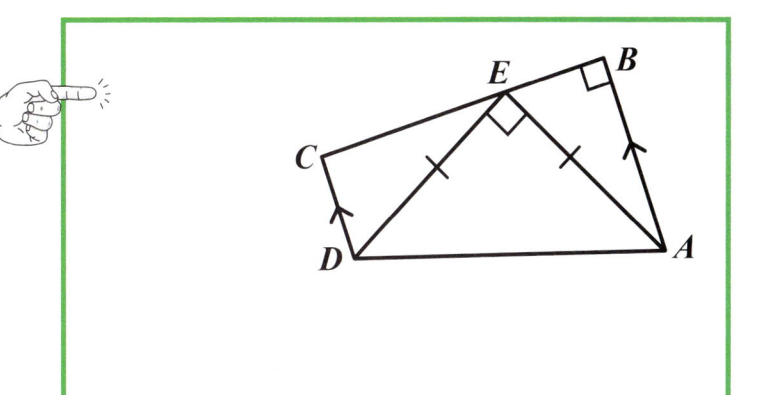

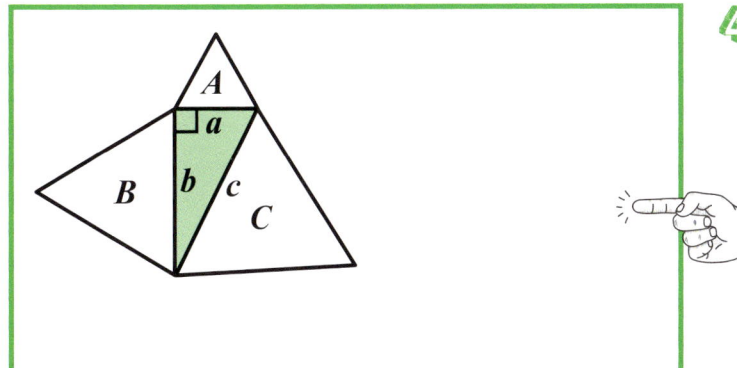

4. A right-angled triangle is formed by joining three equilateral triangles of side lengths a, b and c.

Show that:
   Area of A + Area of B = Area of C

READY?

If one vector is a multiple of another, then the two vectors are parallel. For example, $3a + 2b$ is parallel to $9a + 6b$ since $9a + 6b = 3(3a + 2b)$. Multiplying a vector by -1 results in a vector with the same magnitude, but in the opposite direction.

CHECK-IN

1 Write in terms of p and/or q:

$a =$ $b =$ $c =$ $d =$

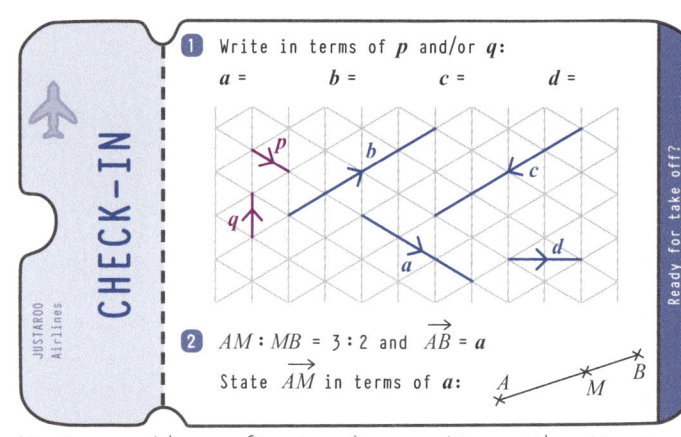

2 $AM : MB = 3 : 2$ and $\overrightarrow{AB} = a$

State $\overrightarrow{AM}$ in terms of a:

---

e.g. 1 $\overrightarrow{PR} = 3a + 2b$

$\overrightarrow{QR} = 4a - b$

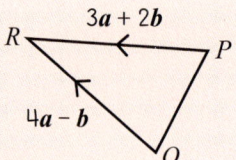

a) Express $\overrightarrow{PQ}$ in terms of a and b.

Find a journey using the known vectors:

$\overrightarrow{PQ} = \overrightarrow{PR} + \overrightarrow{RQ}$

$= 3a + 2b - 4a + b$

$= -a + 3b$

$\overrightarrow{RQ} = -\overrightarrow{QR}$
$= -(4a - b)$
$= -4a + b$

b) The line MN is five times as long as $\overrightarrow{QR}$.

$\overrightarrow{MN}$ is the opposite direction to $\overrightarrow{QR}$.

State $\overrightarrow{MN}$ in terms of a and b.

$\overrightarrow{MN} = -5\overrightarrow{QR} = -5(4a - b) = -20a + 5b$

e.g. 2 $\overrightarrow{XY} = 4a + 3b$ and $\overrightarrow{YZ} = 16a + 12b$

Prove that the points X, Y and Z are in a straight line.

First, show that the vectors are parallel:

$\overrightarrow{YZ} = 16a + 12b = 4(4a + 3b) = 4\overrightarrow{XY}$

Therefore $\overrightarrow{XY}$ and $\overrightarrow{YZ}$ are parallel.

Conclusion And:

The point Y is an endpoint of both vectors so X, Y and Z must lie on a straight line.

Vector problems often involve working with ratios too:

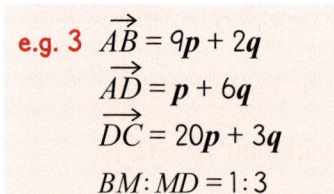

e.g. 3 $\overrightarrow{AB} = 9p + 2q$

$\overrightarrow{AD} = p + 6q$

$\overrightarrow{DC} = 20p + 3q$

$BM : MD = 1 : 3$

Show that $\overrightarrow{AM} = \frac{1}{3}\overrightarrow{AC}$

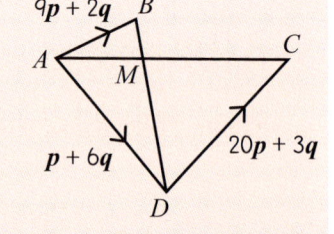

1 Use $BM:MD = 1:3$ to find other vectors to help:

a $\overrightarrow{BD} = \overrightarrow{BA} + \overrightarrow{AD} = -\overrightarrow{AB} + \overrightarrow{AD}$

$= -9p - 2q + p + 6q = -8p + 4q$

b $\overrightarrow{BM} = \frac{1}{4}\overrightarrow{BD}$ *BM is one of four parts of BD*

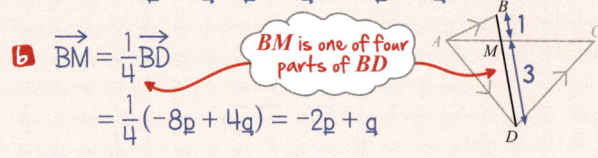

$= \frac{1}{4}(-8p + 4q) = -2p + q$

2 Use these vectors to find $\overrightarrow{AC}$ and $\overrightarrow{AM}$:

a $\overrightarrow{AC} = \overrightarrow{AD} + \overrightarrow{DC}$

$= p + 6q + 20p + 3q = 21p + 9q$

b $\overrightarrow{AM} = \overrightarrow{AB} + \overrightarrow{BM}$

$= 9p + 2q - 2p + q = 7p + 3q$

3 Make a conclusion: $3(7p + 3q) = 21p + 9q$ so $\overrightarrow{AM} = \frac{1}{3}\overrightarrow{AC}$

SET

A.

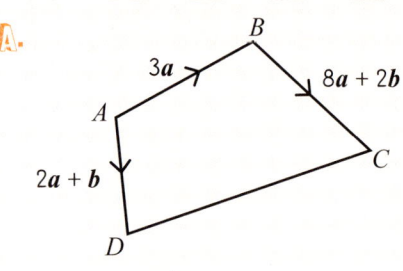

(i) Express $\overrightarrow{CD}$ in terms of a and b.

(ii) The line MN is four times longer than BD.

$\overrightarrow{MN}$ is the opposite direction to $\overrightarrow{BD}$.

Find $\overrightarrow{MN}$ in terms of a and b.

B. $\overrightarrow{PQ} = 2c + 7d$

$\overrightarrow{QR} = 5c + 17.5d$

(i) Explain why P, Q and R lie on a straight line.

(ii) Find the ratio of the length of PQ to the length of QR in the form $n : 1$

C.

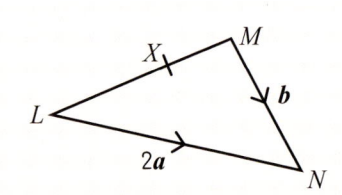

$LX : XM = 5 : 3$

Express $\overrightarrow{NX}$ in terms of a and b.

D.

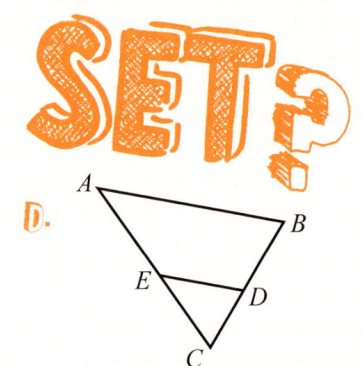

$\overrightarrow{AC} = a$ and $\overrightarrow{BC} = b$

$AE : EC = 3 : 2$

$BD : DC = 3 : 2$

Use vectors to prove that AB is parallel to DE.

114

GO! VECTORS 1

1. *ABCD* is a rhombus.

$\vec{AD} = \mathbf{a}$ and $\vec{AB} = \mathbf{b}$

M is the midpoint of the diagonal *BD*.

The diagonal *AC* also passes through *M*.

Prove that *M* is the midpoint of *AC*.

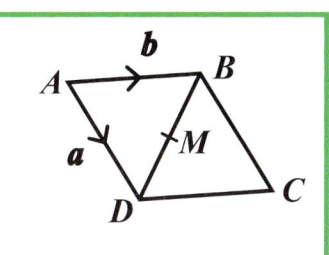

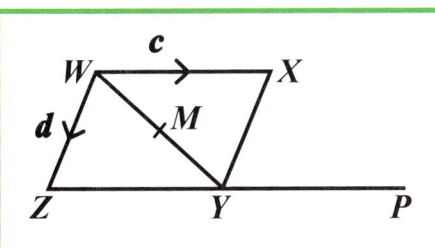

2. *WXYZ* is a parallelogram.

$\vec{WX} = \mathbf{c}$ and $\vec{WZ} = \mathbf{d}$

$ZP = 2ZY$

M and *N* are the midpoints of *WY* and *WP* respectively.

Prove that $\vec{MN}$ is parallel to $\vec{ZP}$.

3. *PSY* and *QRS* are triangles.

Y lies on *RS* such that $SY : YR = 4 : 1$
X lies on *SQ* such that $SX : XQ = 1 : 2$
X is the midpoint of *PY*.

$\vec{SR} = \mathbf{a}$ and $\vec{SQ} = \mathbf{b}$

$\vec{SP}$ can be written in the form $k(-6\mathbf{a} + 5\mathbf{b})$

Find the value of *k*.

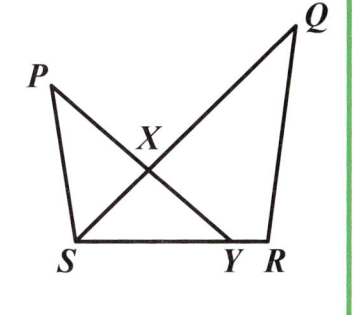

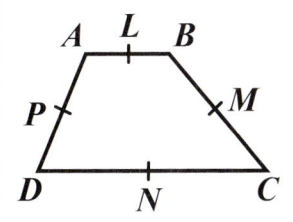

4. *ABCD* is a trapezium.

L, *M*, *N* and *P* are the midpoints of *AB*, *BC*, *CD* and *AD* respectively.

$\vec{AP} = \mathbf{a} \quad \vec{BM} = \mathbf{b} \quad \vec{CN} = \mathbf{c}$

Prove that *LMNP* is a parallelogram.

READY?

If three or more points are in a straight line, then the points are **collinear**. We can use vectors based on these points to solve complex problems.

CHECK-IN

1. Find the value of x if $2x : 7 = x - 3 : 5$
2. Find the value of k if $k : k + 2 = 0.5 : 3$
3. N is the midpoint of BC
   $AM : MB = 1 : 2$
   $\vec{BC} = 12b$
   State $\vec{MN}$ in terms of a of b.

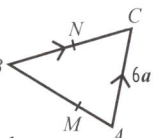

---

e.g. 1
QRT is a triangle.
$\vec{RT} = 3a$
$\vec{ST} = 2a + 4b$
$\vec{PR} = -5a + 6b$

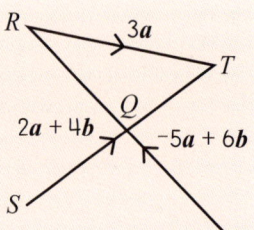

PQR and SQT are straight lines.
Work out the ratio $PQ : QR$.

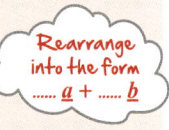

Work with $\vec{QR}$ first as it can be used to form an equation involving other vectors:

1 Write $\vec{QR}$ in terms of $\vec{PR}$:
(k is a fraction we need to find)
$\vec{QR} = k\vec{PR}$ so $\vec{QR} = k(-5a + 6b)$

2 Use $\vec{QR}$ as part of a journey for $\vec{QT}$:
$\vec{QT} = \vec{QR} + \vec{RT} = k(-5a + 6b) + 3a$
$= -5ka + 6kb + 3a$
$= (-5k + 3)a + 6kb$

Rearrange into the form ...... a + ... b

3 Find k by using the fact that $\vec{QT}$ and $\vec{ST}$ must be parallel: 
$\vec{ST} = 2a + 4b$ and $\vec{QT} = (-5k + 3)a + 6kb$
(×2)   (×2)
Spot this
Therefore $2(-5k + 3) = 6k$
$\rightarrow -10k + 6 = 6k \rightarrow 6 = 16k \rightarrow k = \frac{6}{16} = \frac{3}{8}$

4 Make a conclusion: $\vec{QR} = \frac{3}{8}\vec{PR}$ so $PQ : QR = 5 : 3$

---

e.g. 2
ABC is a triangle.
$AN : NB = 3 : 1$
M is the midpoint of AC.
$\vec{AB} = 20b$ and $\vec{AC} = 4a$
Show that $\vec{XN} = -\frac{4}{5}a + 3b$.

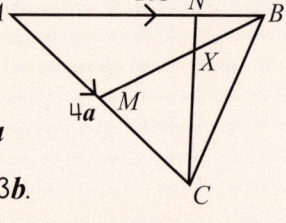

Start by working out the more obvious vector journeys:

$\vec{BC} = \vec{BA} + \vec{AC} = -20b + 4a$
$\vec{MB} = \vec{MA} + \vec{AB} = -2a + 20b$
$\vec{CN} = \vec{CA} + \vec{AN} = -4a + 15b$

$\vec{MA} = -\vec{AM} = -\frac{1}{2} \times 4a = -2a$

Using $AN : NB = 3 : 1$
$\vec{AN} = \frac{3}{4}\vec{AB} = \frac{3}{4} \times 20b = 15b$

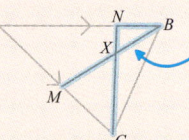

Choose part of the diagram to work with

1 Write $\vec{XN}$ in terms of $\vec{CN}$:
$\vec{XN} = k\vec{CN}$ so $\vec{XN} = k(-4a + 15b)$

2 Use $\vec{XN}$ as part of a journey for $\vec{XB}$:
$\vec{XB} = \vec{XN} + \vec{NB} = k(-4a + 15b) + 5b$
$= -4ka + 15kb + 5b$
$= -4ka + (15k + 5)b$

($\frac{1}{4}\vec{AB}$)

3 Find k by using the fact that $\vec{XB}$ and $\vec{MB}$ must be parallel:
$\vec{MB} = -2a + 20b$ and $\vec{XB} = -4ka + (15k + 5)b$
(×-10)   (×-10)
So $40k = 15k + 5 \rightarrow 25k = 5 \rightarrow k = \frac{1}{5}$

4 Make a conclusion: $\vec{XN} = \frac{1}{5}\vec{CN} = \frac{1}{5}(-4a + 15b) = -\frac{4}{5}a + 3b$

SET?

A.
PQR is a triangle.
A is the midpoint of PQ.
$QB : BR = 1 : 2$
BP and AR intersect at the point X.
$\vec{PQ} = 8a$ and $\vec{QR} = 12b$
$\vec{AX} = k(a + 3b)$.

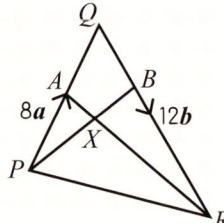

Find the value of k.

B.
$ABCD$ is a parallelogram.
MNP and DCP are straight lines.
N is the midpoint of BD.
$AM : MD = 1 : 2$
$\vec{AB} = 2a$ and $\vec{AD} = 6b$

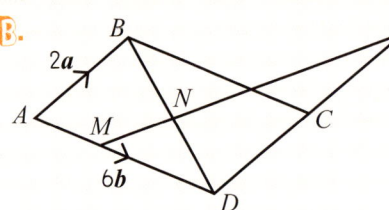

Show that $\vec{MN} = \frac{1}{4}\vec{MP}$.

VECTORS 2

1. PQR and PSY are triangles.

PS = 2PR and Y is the midpoint of PQ.

$\overrightarrow{PQ} = 10a$ $\overrightarrow{PR} = 15b$

$\overrightarrow{QX} = k(2a - 3b)$

Find the value of k.

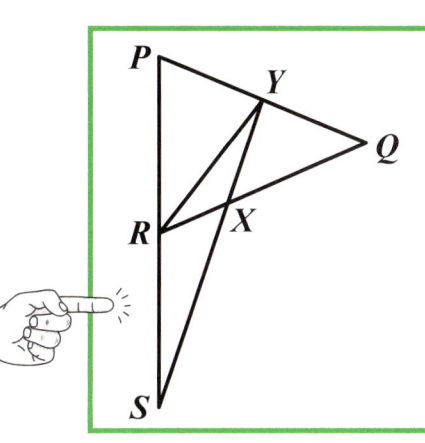

2. ABCDEF is a regular hexagon.

$\overrightarrow{AB} = a$ $\overrightarrow{BC} = b$ $\overrightarrow{CD} = b - a$

M and N are the midpoints of BN and DE respectively.

$\overrightarrow{CM} = ma$

Find the value of m.

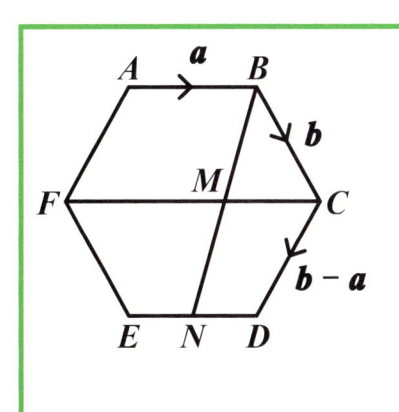

3. Three equilateral triangles are joined to make a trapezium. The lines AD and PC intersect at the point Q.

$\overrightarrow{AB} = a$ $\overrightarrow{AE} = 2b$

AE = 2AP

Find the value of n if $PQ : PC = 1 : n$.

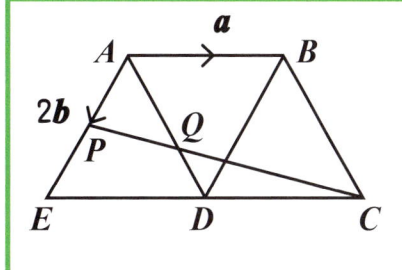

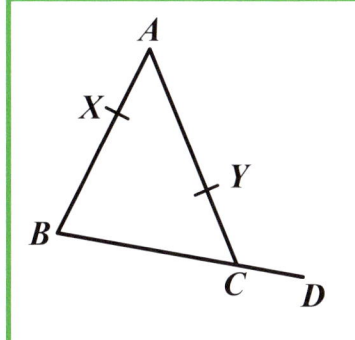

4. AXB, AYC and BCD are straight lines.

X lies on AB such that AX : XB = 1 : 3
Y lies on AC such that AY : YC = 3 : 1

$\overrightarrow{AC} = 8a$ and $\overrightarrow{BC} = 20b$

$\overrightarrow{CD} = kb$

The points X, Y and D lie on a straight line. Find the value of k.

READY?

Tree diagrams, two-way tables and Venn diagrams can be used to sort information. For example, we might know that in a group of 90 people:
- 12 have both a pet dog (D) and a pet cat (C)
- 25 have a dog but do not have a cat
- 43 have neither a dog nor a cat

In a Venn diagram …

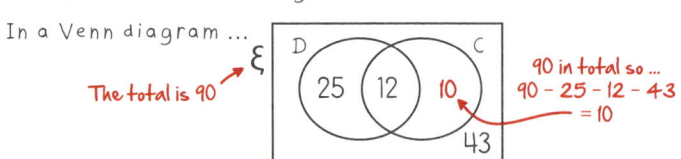

The total is 90 → ξ, D: 25, D∩C: 12, C: 10, outside: 43

90 in total so …
90 − 25 − 12 − 43 = 10

This can help to solve problems such as finding the probability that a person in the group has a cat, but does not have a dog: $\frac{\text{No. of required outcomes}}{\text{Total no. of outcomes}} = \frac{10}{90} \left(= \frac{1}{9}\right)$

We can solve **conditional probability** problems too. These involve only considering some of the possible outcomes:

e.g. 1
The Venn diagram shows information about the members of a cycling club (C), a swimming club (S) and a running club (R) in a town.

A randomly chosen person is a member of both the swimming club and the running club.

What is the probability that the person is also a member of the cycling club?

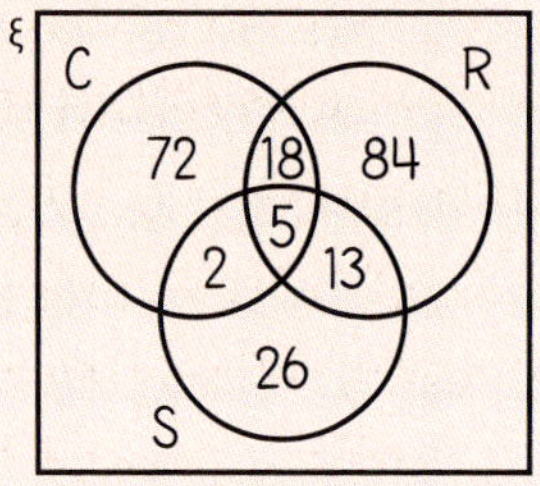

Venn diagram values: C only 72, C∩S 2, C∩R 18, C∩S∩R 5, R only 84, S∩R 13, S only 26

1 Restrict the diagram using the given condition (the person is in both the swimming and running clubs)

2 Use the golden rule:

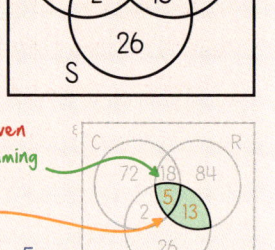

$\frac{\text{Number of required outcomes}}{\text{Total number of outcomes}} = \frac{5}{5+13} = \frac{5}{18}$

CHECK-IN

1 Pupils in a school can choose either PE, RE or Art at the start of Key Stage 4. Complete the frequency tree and two-way table to show the same information:

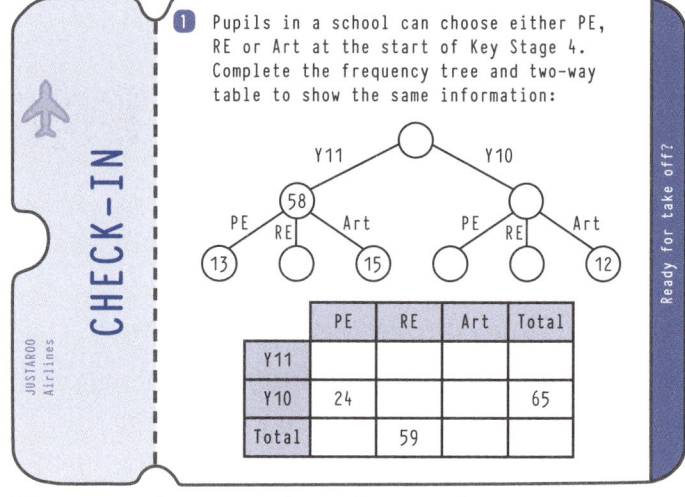

Tree: Total → Y11 (58), Y10; Y11 → PE (13), RE, Art (15); Y10 → PE, RE, Art (12)

	PE	RE	Art	Total
Y11				
Y10	24			65
Total		59		

We can use frequency tree diagrams too …

e.g. 2
A cinema has three screens: A, B and C.

The frequency tree shows the number of tickets sold for adults and children one Saturday afternoon.

One of the tickets is chosen at random to win a prize draw. An adult ticket wins the draw. What is the probability that the winner was from Screen B?

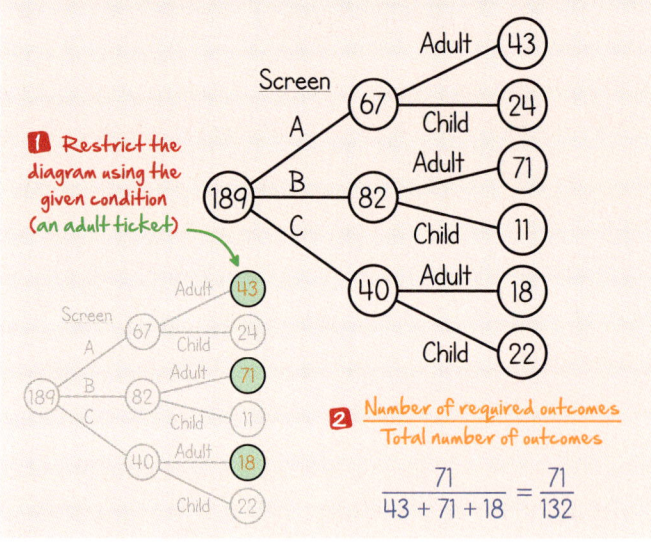

Tree: 189 → A (67) → Adult 43, Child 24; B (82) → Adult 71, Child 11; C (40) → Adult 18, Child 22

1 Restrict the diagram using the given condition (an adult ticket)

2 $\frac{\text{Number of required outcomes}}{\text{Total number of outcomes}}$

$\frac{71}{43 + 71 + 18} = \frac{71}{132}$

SET?

A. 400 people are surveyed. The frequency tree shows information about some of their responses:

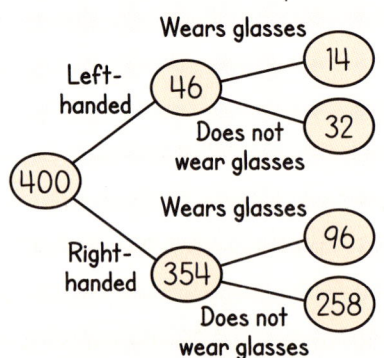

Tree: 400 → Left-handed (46) → Wears glasses 14, Does not wear glasses 32; Right-handed (354) → Wears glasses 96, Does not wear glasses 258

A randomly chosen person wears glasses. What is the probablity that they are left-handed?

B. The Venn diagram shows information about the subjects being studied by some students at a sixth form college:

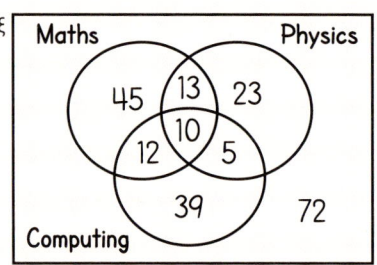

Venn: Maths only 45, Maths∩Physics 13, Physics only 23, Maths∩Computing 12, Maths∩Physics∩Computing 10, Physics∩Computing 5, Computing only 39, outside 72

(i) A student studies physics. What is the probability that they also study maths?

(ii) A student studies maths. What is the probability that they also study both physics and computing?

C. The two-way table gives information about the workers in a company:

	Male	Female	Total
Finance	14		37
Sales	23	12	
Ops			15
Total		44	

(i) Complete the two-way table.

(ii) A male worker is chosen at random. What is the probablity that they work in ops?

GO! PROBABILITY 1

Ace!

1. A transport survey shows the following information about 62 households:
   - 3 have a diesel car and an electric car
   - 5 have a petrol car and an electric car
   - In total, 15 have an electric car, 34 have a petrol car and 28 have a diesel car
   - All have at least one car and no households have all three types of car

   A household with a diesel car is chosen at random. What is the probability that they also have a petrol car?

2. 1 in every 12 males is colour blind.
   1 in every 200 females is colour blind.

   A scientist is studying a group of 600 males and 600 females. They randomly choose a colour blind person. What is the probability that the person they choose is female?

3. In a sixth form college:
   - P(student studies music) = $\frac{7}{60}$
   - P(student studies music and maths) = $\frac{1}{15}$

   The probability that a student studies neither of these subjects is $\frac{7}{20}$.

   A randomly chosen student studies maths. Find P(this student studies music).

   HINT: How many students could there be?

4. There are 240 pupils in a year group.
   48 of the pupils study GCSE PE and 60 study history. N pupils study neither of these two subjects.

   A randomly chosen pupil studies history. Show that the probability they also study GCSE PE is given by the expression: $\frac{N-132}{60}$

READY?

Tree diagrams can be used to represent a probability problem when there are two (or more) combined events. They can also help to solve more complex problems when some of the information is missing.

e.g. 1 A bag contains grey and blue beads in the ratio 1 : 5
Two beads are taken from the bag at random.*
The probablity that a bead of each colour is chosen is $\frac{2}{7}$
Let n be the number of grey counters.

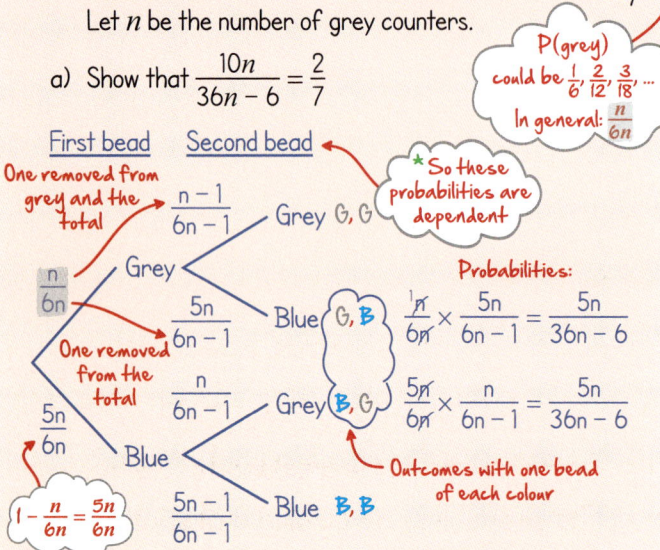

a) Show that $\frac{10n}{36n-6} = \frac{2}{7}$

> P(grey) could be $\frac{1}{6}, \frac{2}{12}, \frac{3}{18}, \ldots$
> In general: $\frac{n}{6n}$

First bead — **Second bead**

One removed from grey and the total: $\frac{n-1}{6n-1}$ Grey G,G

$\frac{n}{6n}$ Grey

$\frac{5n}{6n-1}$ Blue G,B

One removed from the total: $\frac{n}{6n-1}$ Grey B,G

$\frac{5n}{6n}$ Blue

$1 - \frac{n}{6n} = \frac{5n}{6n}$

$\frac{5n-1}{6n-1}$ Blue B,B

*So these probabilities are dependent

Probabilities:
$\frac{n}{6n} \times \frac{5n}{6n-1} = \frac{5n}{36n-6}$

$\frac{5n}{6n} \times \frac{n}{6n-1} = \frac{5n}{36n-6}$

Outcomes with one bead of each colour

$P(G,B \text{ or } B,G) = P(G,B) + P(B,G) = \frac{10n}{36n-6} = \frac{2}{7}$

b) Work out the number of grey beads that were in the bag originally.

$\frac{10n}{36n-6} = \frac{2}{7}$ so $\frac{70n}{36n-6} = 2$ — Multiply both sides by $(36n-6)$

Multiply both sides by 7
so $70n = 72n - 12$
so $12 = 2n$
so $n = 6$

There were six grey beads in the bag

CHECK-IN

A bag contains 13 green cubes and 7 orange cubes. Leah takes a cube at random, keeps it, and then takes another cube.

1. Complete the tree diagram:

First cube — Second cube
...... Green
...... Green
...... Orange
$\frac{7}{20}$ Orange
...... Green
...... Orange

2. Work out the probability that the two cubes are different colours.

---

e.g. 2 There are x blue and 7 red counters in a box.
A counter is picked out at random and placed on a table. Another counter is then picked out at random.

The probability of picking out two blue counters is $\frac{5}{33}$

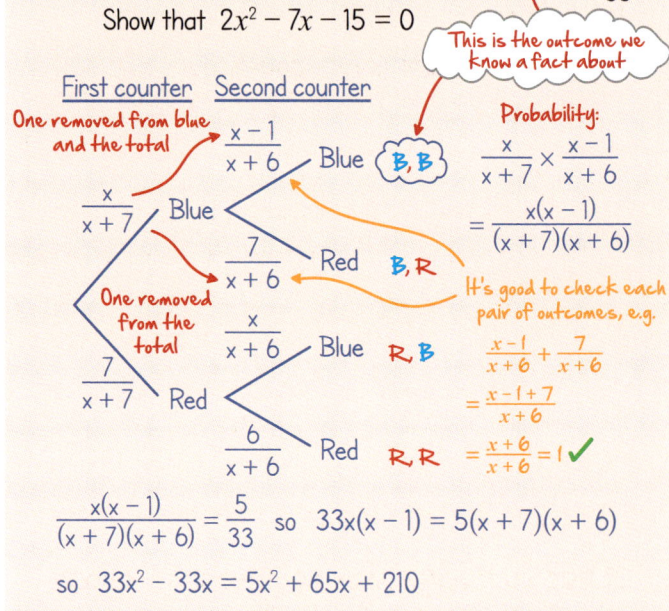

Show that $2x^2 - 7x - 15 = 0$ ← This is the outcome we know a fact about

First counter — **Second counter**

One removed from blue and the total: $\frac{x-1}{x+6}$ Blue B,B

$\frac{x}{x+7}$ Blue

$\frac{7}{x+6}$ Red B,R

One removed from the total: $\frac{x}{x+6}$ Blue R,B

$\frac{7}{x+7}$ Red

$\frac{6}{x+6}$ Red R,R

Probability:
$\frac{x}{x+7} \times \frac{x-1}{x+6}$
$= \frac{x(x-1)}{(x+7)(x+6)}$

It's good to check each pair of outcomes, e.g.
$\frac{x-1}{x+6} + \frac{7}{x+6}$
$= \frac{x-1+7}{x+6}$
$= \frac{x+6}{x+6} = 1$ ✓

$\frac{x(x-1)}{(x+7)(x+6)} = \frac{5}{33}$ so $33x(x-1) = 5(x+7)(x+6)$
so $33x^2 - 33x = 5x^2 + 65x + 210$
so $28x^2 - 98x - 210 = 0 \xrightarrow{\div 14} 2x^2 - 7x - 15 = 0$

A. A box contains red counters and green counters in the ratio 3 : 2.
Two counters are taken at random from the box.
$P(\text{choosing two green counters}) = \frac{1}{7}$
The number of green counters is $2n$.

(i) Show that $14(2n-1) = 5(5n-1)$

(ii) Work out the number of green counters that were in the box.

B. A box contains 5 red counters and n green counters.
Two counters are taken at random from the box.
The probability of choosing one counter of each colour is $\frac{10}{21}$
Show that $n^2 - 12n + 20 = 0$

SET?

C. A bag contains n beads. Six of the beads are red.
Three beads are taken from the bag at random.
Show that the probability that none of the beads are red is:

$$\frac{n^3 - 21n^2 + 146n - 336}{n^3 - 3n^2 + 2n}$$

GO! PROBABILITY 2

Lush!

1. A box contains x plastic cubes. Two cubes are orange and the rest are blue. A cube is chosen at random, and then replaced. A cube is then chosen from the box again.

   a) Show that P(two blue cubes) = $\dfrac{x^2 - 4x + 4}{x^2}$

   b) Dom is told that P(two blue cubes) = $\dfrac{64}{81}$.
   He writes: $x^2 = 81$, so $x = 9$
   Explain why Dom is not correct.

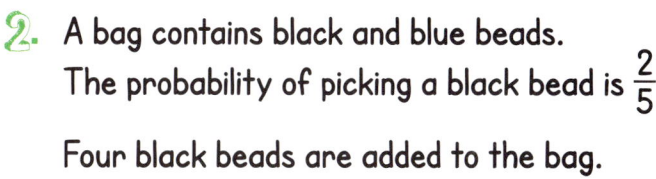

2. A bag contains black and blue beads. The probability of picking a black bead is $\dfrac{2}{5}$

   Four black beads are added to the bag. The probability of picking a black bead increases to 0.5

   How many blue beads are in the bag?

3. A bag contains yellow counters and green counters. The ratio of yellow counters to green counters is 5 : 3.

   A counter is taken at random from the bag and placed on a desk. Another counter is then taken from the bag.

   P(two yellow counters) = $\dfrac{35}{92}$

   How many green counters were in the bag before any were chosen?

4. Ida has 6 black pens and n red pens in her pencil case.

   If Ida takes two pens at random from her pencil case, the probability that both pens are red is $\dfrac{2}{15}$.

   Find the total number of black and red pens Ida had in her pencil case.

READY?

The median is the value 'half way through' a set of ordered data. For example, 21 is the median here:

13 17 18 19 19 **21** 23 24 26 28 31

The lower quartile (LQ) is another measure of location. It is one quarter of the way through the ordered data. With a small set of data, the LQ can be found as follows:

13 17 **18** 19 19 **21** 23 24 26 28 31

18 is the median of the remaining lower numbers

> The lower quartile of n numbers* is the $\frac{1}{4} \times (n+1)^{th}$ number

The upper quartile (UQ) is three quarters of the way through the ordered data. It is found in a similar way:

13 17 **18** 19 19 **21** 23 24 **26** 28 31

The median of the remaining upper numbers is 26

> The upper quartile of n numbers* is the $\frac{3}{4} \times (n+1)^{th}$ number

*The numbers must be in order

e.g. 1 The heights of 11 plants are measured in centimetres:

37 42 32 35 41 38 37 45 36 33 41

Find the upper quartile for the set of data.

1 Write the data in order, starting with the smallest:

32 33 35 36 37 (37) 38 41 41 42 45

2 Identify the median

3 Find the median of the remaining upper numbers

The upper quartile is 41

OR Use $\frac{3}{4} \times (n+1) \to n = 11$, so $\frac{3}{4} \times (11+1) = \frac{3}{4} \times 12 = 9$

The 9th number is 41 (when the numbers are in order)

CHECK-IN

1 Find the median for each set of data:

104 105 108 113 114 121 124

32.2 29.1 30.3 19 23.4 21 24.8

76 34 56 23 58 62 48 72

The interquartile range (IQR) is a way of measuring the spread of the data. It is worked out by finding the difference between the upper and lower quartiles ...

e.g. 2 The weights of some tomatoes are measured in grams:

23 24 27 28 30 30 33 35 36 38 39

Work out the interquartile range for the set of data.

1 Check the data is in order; identify the median and the quartiles:

23 24 **27** 28 30 (30) 33 35 **36** 38 39

LQ = 27   Median = 30   UQ = 36

2 Work out the difference between the upper and lower quartiles:

Interquartile range = UQ − LQ = 36 − 27 = 9

Sets of data can be compared by using an average and a measure of spread. We often use the median and the IQR or range to make this comparison:

e.g. 3 The table summarises the test results of two classes:

	Median	Highest	Lowest	LQ	UQ
Class A	52	86	31	42	68
Class B	59	83	19	25	63

Compare the distribution of results for the classes.

IQR for Class A: UQ − LQ = 68 − 42 = 26
IQR for Class B: UQ − LQ = 63 − 25 = 38

On average, Class B had higher test results with a median of 59. The results from Class A were more consistent as their interquartile range of 26 was lower.

We could use the range instead of the IQR

A. Work out the lower and upper quartiles for each set of data:

(i) 104 105 108 113 114 121 124

(ii) 32.2 29.1 30.3 19 23.4 21 24.8

(iii)
23	42	27	35	31
46	37	36	28	39
35	40	33	41	28

(iv) 16 22 24 25 27 27 30 32

B. Work out the interquartile range for each set of data in **A**.

SET?

C. Here is some summarised data about the heights of pupils in two classes. The measurements are in centimetres.

	Highest	Lowest	Median	LQ	UQ
Class A	172	147	162	157	168
Class B	176	148	161	152	165

Compare the distribution of heights in the classes.

GO! STATISTICS

1. Vic is asked to compare the distribution of test results for the two classes shown. She says:

Class	Median	LQ	UQ	Lowest	Highest
A	63	47	72	24	95
B	63	45	68	28	87

   "Class A scored better results on the test as their highest value and upper quartile are higher than Class B"

   Do you agree with Vic? Explain why.

2. Zain is asked to work out the interquartile range for this set of data:

58	45	69	49	73	51	74	70
63	59	61	72	51	64	55	

   He writes:

   IQR = UQ − LQ = 74 − 45 = 29

   Explain the mistake that Zain has made.

3. The table shows data about the lengths of phonecalls in a call centre.

   a) Which group contains the median?
   b) In which group does the upper quartile lie?
   c) Explain why the interquartile range cannot be worked out.

Time (t seconds)	Frequency
$0 \leq t < 30$	17
$30 \leq t < 60$	23
$60 \leq t < 90$	35
$90 \leq t < 120$	10
$120 \leq t < 150$	2

   ```
       Group B        Group A
           9 6 | 14 | 8
         7 5 2 | 15 | 3 4 6 6
   9 8 8 4 3 0 | 16 | 2 2 4 5 8
         9 7 4 1 | 17 | 0 1 6
                 | 18 | 1 3
   ```

4. The chart shows information about the armspans of two groups of people. The key for the diagram is:

   Group A   Group B
   14 | 8 = 148 cm   6 | 14 = 146 cm

   Compare the distribution of the data for the two groups.

READY?

A **box plot** (or box and whisker diagram) is a chart which shows information about the location of key values in a set of data:

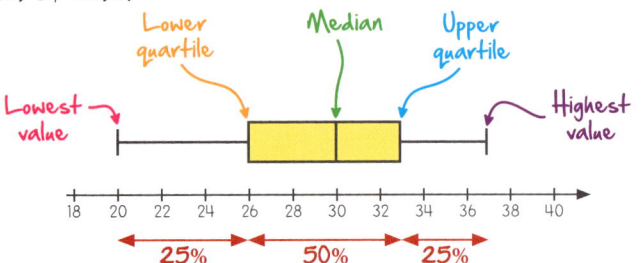

- Lowest value
- Lower quartile
- Median
- Upper quartile
- Highest value

Scale: 18, 20, 22, 24, 26, 28, 30, 32, 34, 36, 38, 40

25% | 50% | 25%

The box in the middle of the chart represents the interquartile range (the spread of the middle 50%). The 'whiskers' each represent the spread of 25% of the data.

e.g. 1
The handspans (in cm) of a group of 15 people are:

14 15 16 16 17 18 18 19 20 20 21 21 23 23 24

Draw a box plot for this data.

1 Identify the lowest, lower quartile, median, upper quartile and highest:

~~14~~ ~~15~~ ~~16~~ 16 ~~17~~ ~~18~~ ~~18~~ (19) ~~20~~ ~~20~~ ~~21~~ 21 ~~23~~ ~~23~~ ~~24~~

2 Draw and label a horizontal scale:

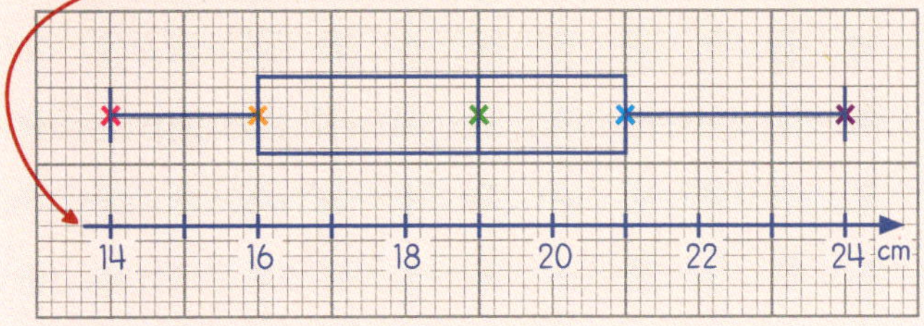

Scale: 14, 16, 18, 20, 22, 24 cm

3 Plot the five key values (14, 16, 19, 21 and 24)

4 Draw in the box and whiskers

CHECK-IN

1 Find the median and interquartile range for the set of data:

| 80 | 65 | 50 | 79 | 52 | 61 | 71 | 72 |
| 58 | 61 | 63 | 83 | 73 | 55 | 59 |

We can use box plots to compare sets of data ...

e.g. 2
The box plot shows information about the diameters of some pumpkins at Farm A.

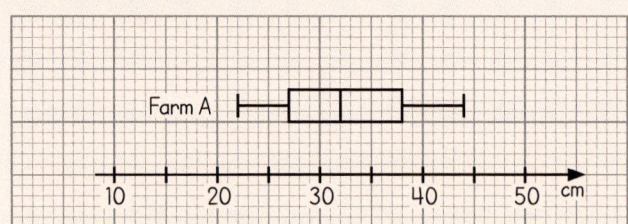

Farm A box plot on scale 10 to 50 cm.

Here is data about the diameters of some pumpkins at Farm B:

33 19 28 30 31 37 25 31 34 26 26

a) Draw a box plot for the data from Farm B.

~~19~~ 25 26 ~~26~~ ~~28~~ (30) ~~31~~ ~~31~~ 33 ~~34~~ ~~37~~

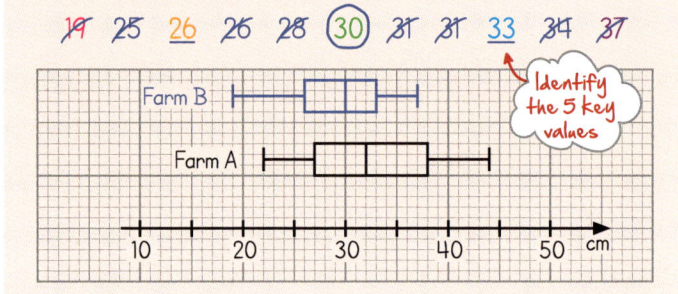

(Identify the 5 key values)

Farm B and Farm A box plots on scale 10 to 50 cm.

b) Compare the distribution of pumpkin sizes for the two farms.

On average, the largest pumpkins are grown at Farm A as the median is larger than for Farm B (32 cm > 30 cm).

The size of pumpkins at Farm B is more consistent as the interquartile range is smaller than for Farm A (7 cm < 11 cm).

SET?

A. The masses, in kilograms, of 15 dogs are shown in the table:

43	28	19	56	31
29	23	12	20	35
50	24	37	21	34

Draw a box plot to represent the data.

B. Here is some information about the number of goals scored by players in two football teams in a season:

	Median	LQ	UQ	Lowest	Highest
Team 1	12	3	17	1	38
Team 2	14	5	15	3	29

(i) Draw a box plot showing the data for each team.
(ii) Compare the distribution of goals scored.

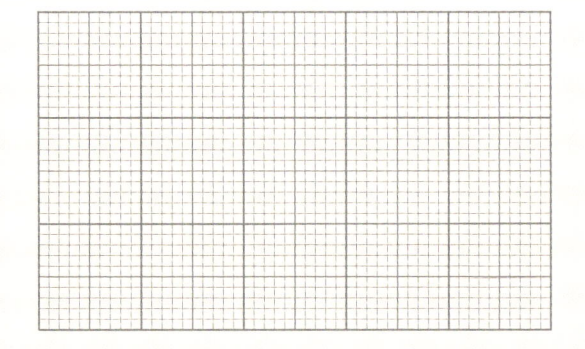

GO! BOX PLOTS

1. The box plot represents a set of 15 values. Write a possible set of 15 values that could be shown by the box plot.

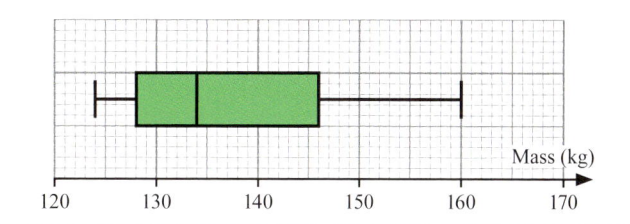

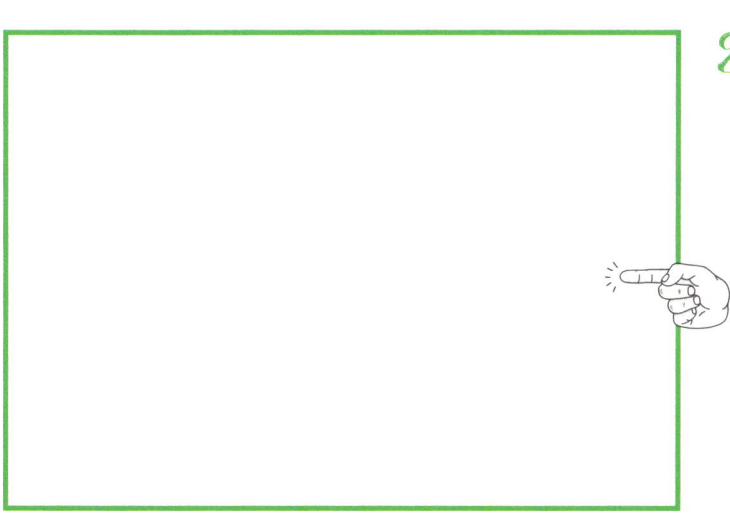

(Box plot: min 124, LQ 129, median 134, UQ 146, max 160, Mass (kg))

2. The box plots show information about the ages of people living in two towns:

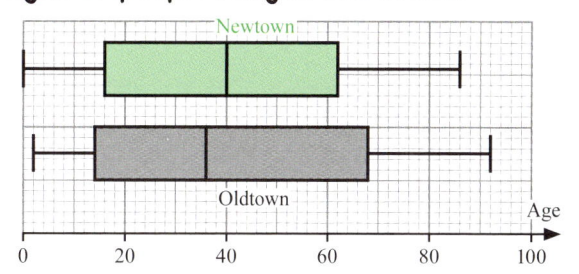

(Box plots: Newtown and Oldtown, Age axis 0 to 100)

Alan thinks that at least half the population is over the age of 40 in both towns. Explain why Alan is not correct.

3. The box plots show information about daily visits to a website at the start of 2024.

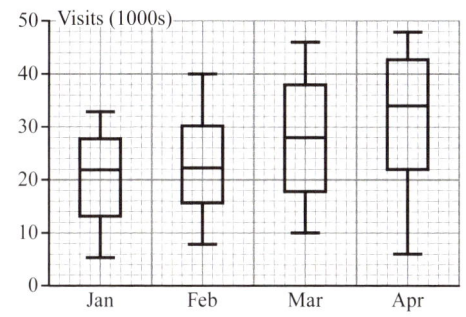

(Box plots for Jan, Feb, Mar, Apr, Visits (1000s))

Describe the pattern of visits over time.

4. The upper quartile of a set of data is 72 cm.
The range is double the interquartile range.
The difference between the upper quartile and the median is two thirds of the interquartile range.
The lowest value is 43 cm.
The interquartile range is 21 cm.
Draw a box plot for the data.

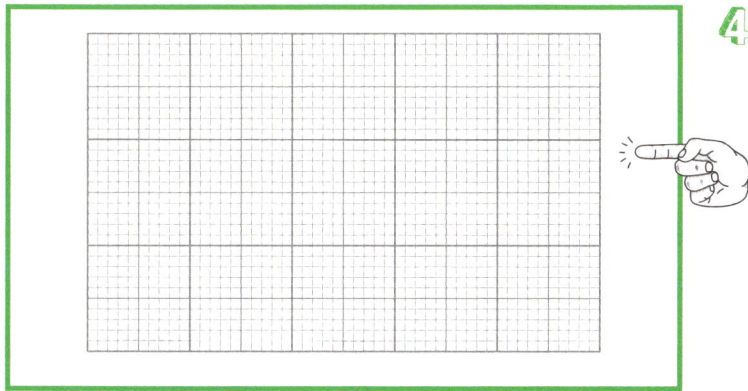

READY?

It is often useful to look at the running totals of data in a grouped frequency table. These totals are also known as the **cumulative frequency** (CF).

e.g. 1a Work out the cumulative frequencies for this data:

Height (h cm)	Frequency	CF
$120 \leq h < 130$	2	2
$130 \leq h < 140$	8	10 (+8)
$140 \leq h < 150$	17	27 (+17)
$150 \leq h < 160$	9	36 (+9)
$160 \leq h < 170$	4	40 (+4)

We can plot a **cumulative frequency diagram** to show these running totals. The points are always plotted at the **upper end** of the intervals:

e.g. 1b Plot a cumulative frequency diagram for the data.

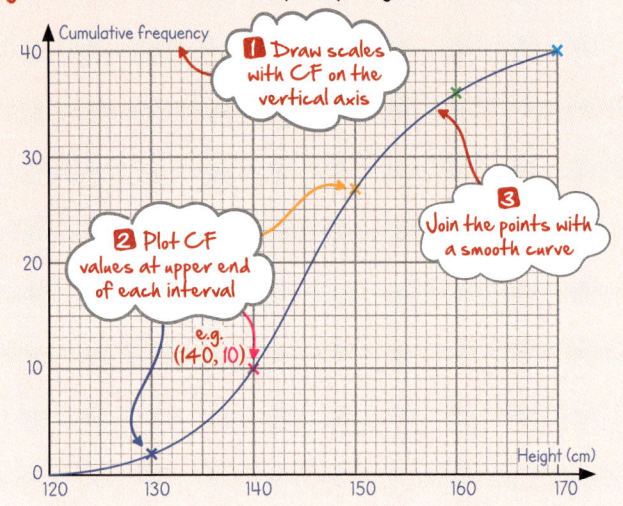

1. Draw scales with CF on the vertical axis
2. Plot CF values at upper end of each interval e.g. (140, 10)
3. Join the points with a smooth curve

CHECK-IN

1 Here is information about the scores in a season of two batters in a cricket team:

	Lowest	Highest	Median	LQ	UQ
Aishah	3	152	46	18	75
Mia	10	98	44	21	67

Compare the distribution of scores for the two batters.

Cumulative frequency diagrams can be used to make estimations more easily than from the frequency table:

e.g. 2 The cumulative frequency diagram shows lengths of journeys to work for staff at a company.

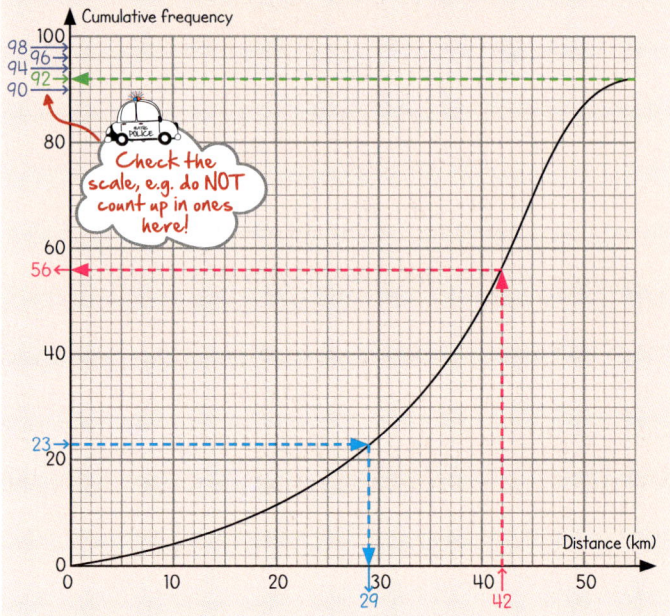

Check the scale, e.g. do NOT count up in ones here!

a) How many staff are in the set of data? **92**
b) Estimate the lower quartile for the data.
   When the number of items in a set of data (n) is large (say $n > 30$), we can use $n \div 4$ to estimate the position of the lower quartile:
   $92 \div 4 = 23$ so the lower quartile = 23rd value = **29 km**
c) Estimate the number of journeys **greater** than 42 km.
   $92 - 56 = 36$ (56 journeys would be **less** than 42 km)

SET?

A. Plot a cumulative frequency diagram for the data in the frequency table.

Height (h cm)	Frequency
$10 < h \leq 12$	4
$12 < h \leq 14$	12
$14 < h \leq 16$	27
$16 < h \leq 18$	21
$18 < h \leq 20$	6

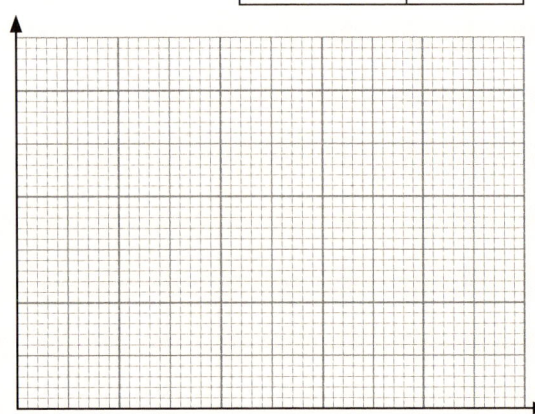

B. The diagram shows information about the lengths of some worms.

(i) Estimate the median length.
(ii) Estimate the upper quartile of the data.
(iii) Estimate the number of worms greater than 10 cm long.

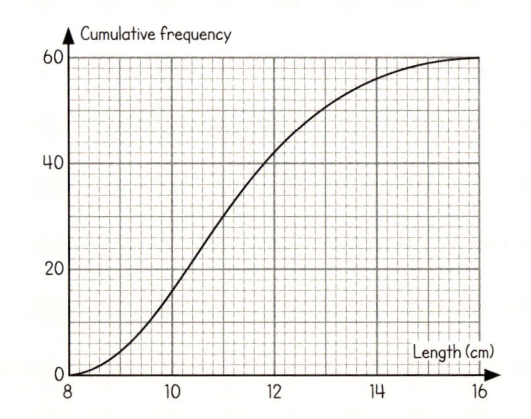

GO! CUMULATIVE FREQUENCY

Top notch!

1. Lois draws a cumulative frequency diagram to represent the information in the table:

Distance (d km)	Frequency
20 < d ≤ 40	4
40 < d ≤ 60	8
60 < d ≤ 80	32
80 < d ≤ 100	16

   Find two mistakes that Lois has made.

2. The graph shows information about how long 240 runners took to complete a 10 km race.

   Compare the distribution of times taken by men and women to finish the race.

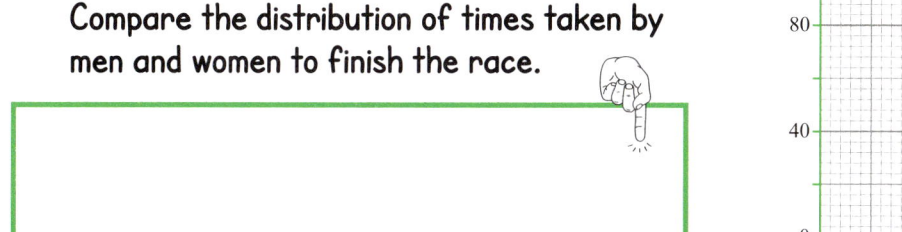

3. The cumulative frequency diagram shows information about the price that 120 people paid for a concert ticket.

   During the concert, one person is chosen at random. What is the probability that this person had paid more than £70 for their ticket?

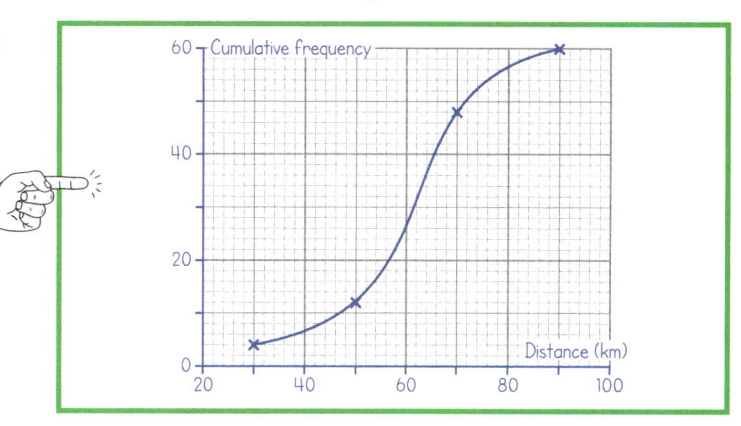

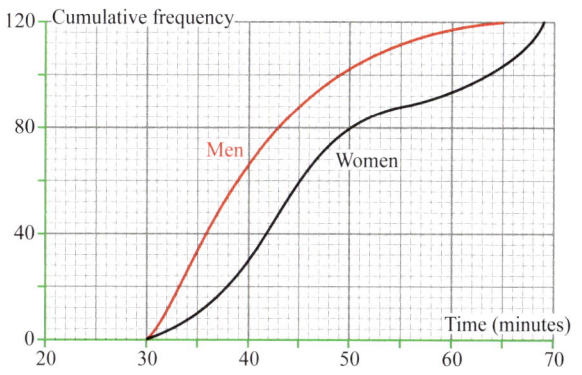

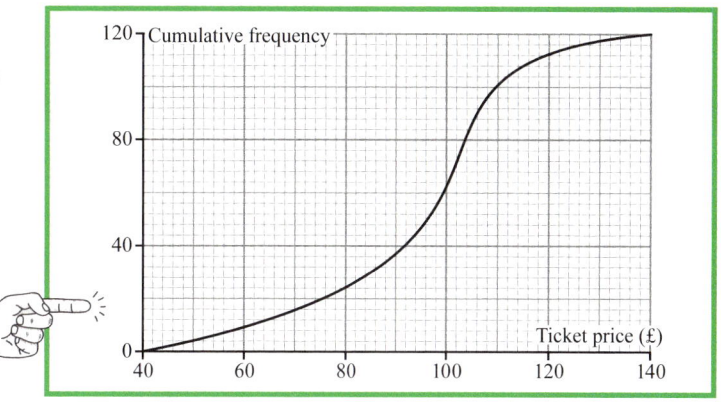

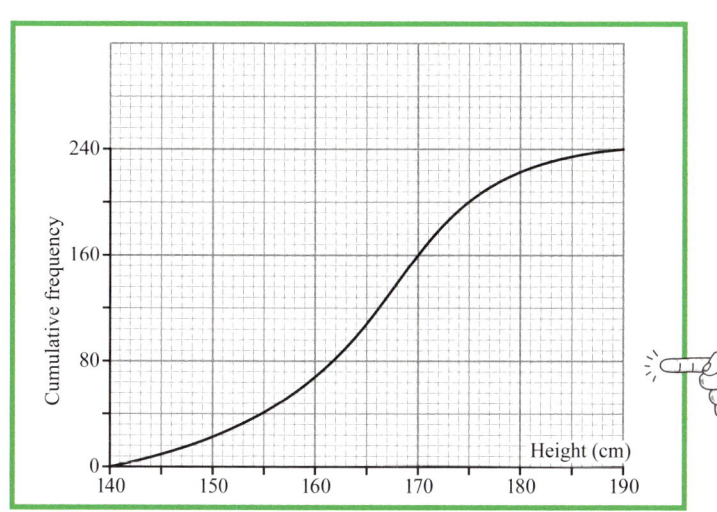

4. Data about the heights of 240 people is shown on the cumulative frequency diagram.

   The shortest height is 142 cm.
   The tallest height is 184 cm.

   Use this information and the diagram to construct a box plot for the 240 people.

READY?

Histograms are a type of frequency diagram that can be used for grouped continuous data. Each group is represented by a rectangular bar. The area of the bar is proportional to the frequency of the group. If the groups are equal widths this is straightforward:

Height (h cm)	Frequency
$50 \leq h < 60$	4
$60 \leq h < 70$	8
$70 \leq h < 80$	16
$80 \leq h < 90$	12
$90 \leq h < 100$	2

← All group widths of 10

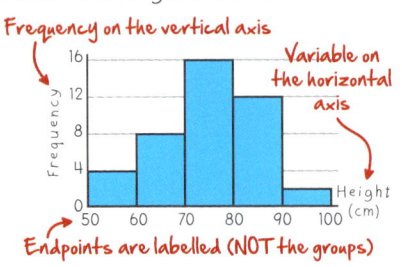

→ Frequency on the vertical axis
→ Variable on the horizontal axis
→ Endpoints are labelled (NOT the groups)

If the groups are unequal, we use **frequency density** to keep the area of the bars proportional to the frequency:

Frequency density = frequency ÷ group width

e.g. 1a
The table shows information about the masses of some dogs. Work out the frequency density for each group.

Mass (m kg)	Frequency	Group width	Frequency density
$30 \leq m < 50$	8	20	0.4
$50 \leq m < 60$	9	10	0.9
$60 \leq m < 65$	14	5	2.8
$65 \leq m < 70$	10	5	2
$70 \leq m < 80$	6	10	0.6

E.g. for the $30 \leq m < 50$ group: ① Group width = 50 − 30 = 20
② Frequency density = 8 ÷ 20 = 0.4

CHECK-IN

1. Which group contains the median?
2. In which group does the upper quartile lie?

Mass (m kg)	Frequency
$0 \leq m < 10$	8
$10 \leq m < 20$	15
$20 \leq m < 30$	42
$30 \leq m < 40$	22
$40 \leq m < 50$	13

Ready for take off?

e.g. 1b
Plot a histogram to show the information about the masses of dogs.

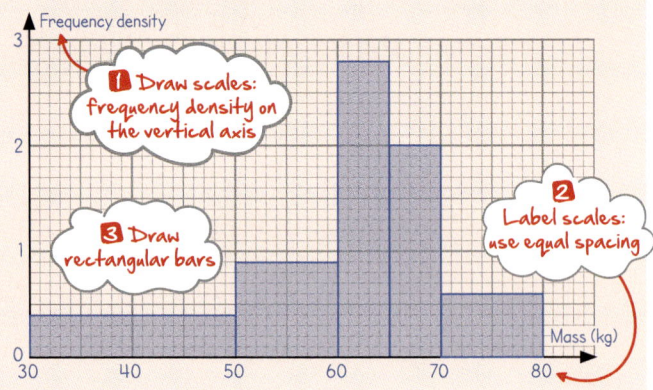

① Draw scales: frequency density on the vertical axis
② Label scales: use equal spacing
③ Draw rectangular bars

As: frequency density = frequency ÷ group width
then: frequency density × group width = frequency
This means we can solve problems using the fact that:
Area of bar (length × width) = frequency

e.g. 1c
Estimate the number of dogs with a mass between 35 kilograms and 60 kilograms.

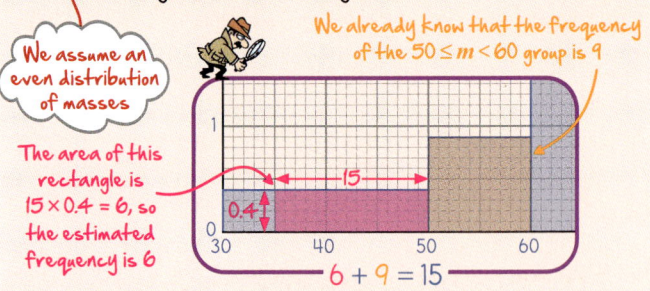

We assume an even distribution of masses

We already know that the frequency of the $50 \leq m < 60$ group is 9

The area of this rectangle is $15 \times 0.4 = 6$, so the estimated frequency is 6

$6 + 9 = 15$

The number of dogs between 35 kg and 60 kg is about 15

A. Here is data about the area of some plots of land. Plot a histogram for the data.

Area (A m²)	Frequency
$50 < A \leq 70$	4
$70 < A \leq 80$	15
$80 < A \leq 90$	13
$90 < A \leq 110$	14
$110 < A \leq 140$	6

B. The histogram shows information about the length of lessons in different schools.

(i) Work out the number of schools with lessons between 50 and 60 minutes long.

(ii) Estimate the number of schools with lessons longer than 65 minutes.

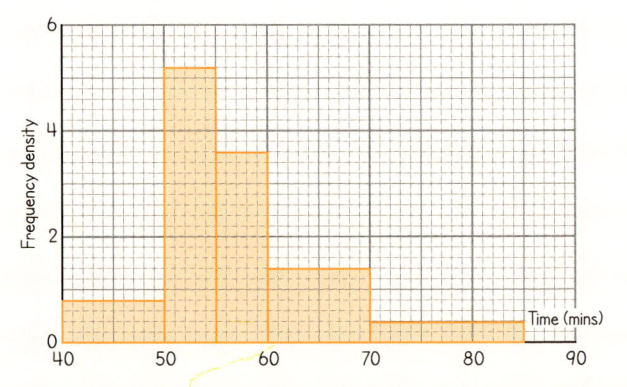

SET?

GO! HISTOGRAMS

1. Tick the chart that it not a histogram.

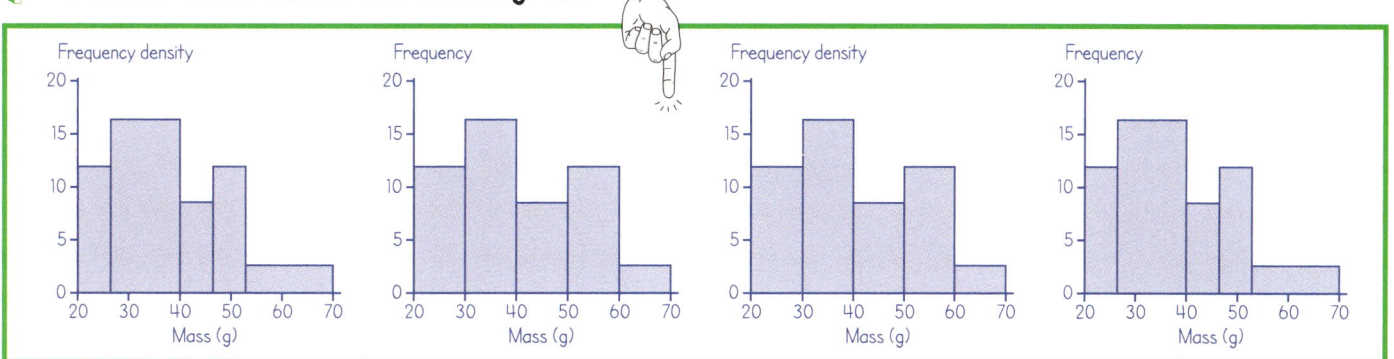

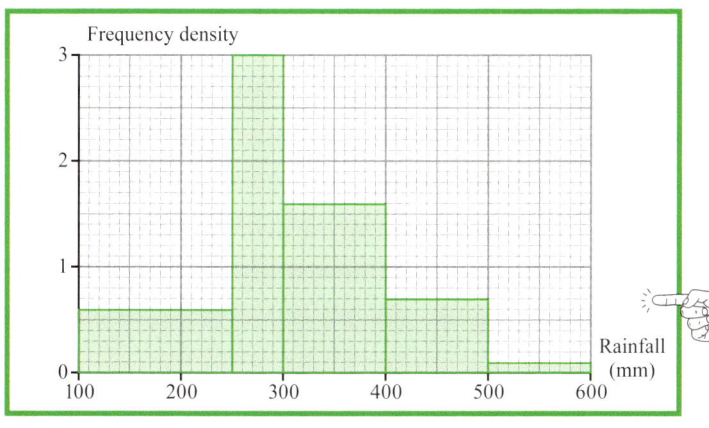

2. The histogram shows information about annual rainfall in different locations around the UK.
   a) Show that an estimate for the lower quartile of the data is 260 millimetres.
   b) Work out an estimate for the interquartile range of the data.

3. The histogram shows information about the recorded speeds of cars on a stretch of road.
   a) Calculate an estimate of the median speed.
   b) Calculate an estimate for the mean of the data.

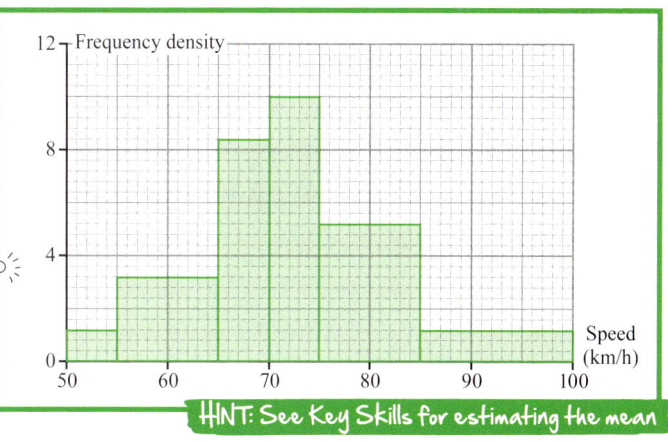

   HINT: See Key Skills for estimating the mean

4. The frequency table and histogram show information about the same set of data.

Mass (m grams)	Frequency
$400 < m \leq 500$	12
$500 < m \leq 550$	17
$550 < m \leq 600$	
$600 < m \leq 700$	10
$700 < m \leq 850$	

   Complete the histogram.

READY?

A population is the whole group of people or items being studied. A sample is a smaller part of a population. It is sometimes possible to use a sample to estimate the size of a population. For example:

Here is a large pile of counters.

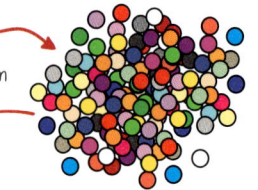

Choose 20 counters and mark them with a cross ...

... and then mix them back in.

Now take a random sample (of any size) and write the proportion of them that are marked with a cross.

This sample size = 30 → $\frac{5}{30}$ of counters have a cross on them

We estimate the size of the population (N) by finding the missing value here:

Proportion with a cross in the sample → $\frac{5}{30} = \frac{20}{N}$ ← Proportion with a cross in the population

We assume that the proportions are equal

$\frac{5}{30} = \frac{20}{120}$ (×4)

So an estimate for the size of the population is 120

This capture-recapture method is very useful when it is not possible to count the population. It is often used to estimate populations of wild animals.

CHECK-IN

1. Find the missing values:

$\frac{80}{a} = \frac{16}{75}$ $\frac{4}{50} = \frac{b}{200}$

$\frac{8}{20} = \frac{70}{c}$ $\frac{d}{15} = \frac{24}{40}$

e.g. 1 Brian wants to know how many bats are living in a cave.

He catches 60 bats and tags them. The bats are returned to the cave.

Ten days later Brian returns to the cave and catches 100 bats. 12 of these bats have a tag on them.

a) Use Brian's results to estimate the population of bats.

1. Write an equation using the proportion of tagged bats in the sample and the population: $\frac{12}{100} = \frac{60}{N}$

2. Find the missing value in the equivalent fractions: $\frac{12}{100} = \frac{60}{500}$ (×5)

An estimate for the population of bats is 500

b) Write down one assumption that you have made.

None of the tags fell off

Sometimes the numbers are not integers ...

e.g. 2 Michaela catches and rings 50 of the birds on an island before releasing them. One week later she catches 40 birds and notices that 13 of them are ringed.

a) Use these results to estimate the population of birds on the island.

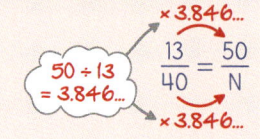

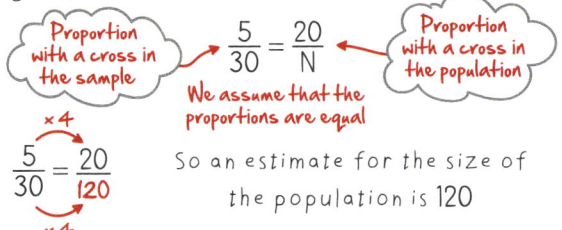

$50 \div 13 = 3.846...$ $\frac{13}{40} = \frac{50}{N}$ ×3.846...

$N = 40 \times 3.846...$
$= 153.846...$
$= 154$ (nearest whole number)

An estimate for the population of birds is 154

b) Write down one assumption that you have made.

The ringed birds are randomly spread across the island

SET?

A. Ruth catches and marks 40 butterflies.

Two weeks later Ruth returns to the same location and catches 50 butterflies. She notices that 10 of them are marked.

(i) Use Ruth's results to estimate the size of the population of butterflies.

(ii) Write down one assumption that you have made.

B. Ewan finds 24 snails in his garden. He writes a number on each of their shells and releases them back into the garden.

Ten days later Ewan finds 20 snails in his garden. 5 of them have a number on their shell.

(i) Estimate the number of snails in Ewan's garden.

(ii) Write down two assumptions that you have made.

C. Hamdi has a large jar full of sweets. She removes 50 of the sweets, puts a mark on them and then mixes them back in the jar.

The next day Hamdi takes out 80 of the sweets and notices that 7 of them are marked.

(i) Estimate the number of sweets in the jar.

(ii) Write down one assumption that you have made.

GO! CAPTURE-RECAPTURE

Impressive!

1. Heather is a beekeeper.

   She catches 60 bees from a hive and notices that 5 of them have been marked.

   Two weeks earlier she had caught and marked 80 of the bees.

   a) Estimate the number bees in the hive.
   b) Write down any assumptions you have made.

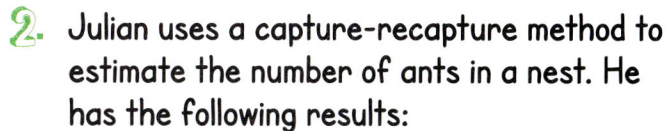

2. Julian uses a capture-recapture method to estimate the number of ants in a nest. He has the following results:
   - First visit: 100 ants caught and marked
   - Second visit: 500 ants caught, of which 12 are marked

   Julian estimates the population of ants by working out:

   $$25 \times 500 \div 3 \approx 4170$$

   Do you agree with Julian? Explain why.

3. Alan, Bob and Caz want to estimate the size of the population of fish in a lake. They catch 100 fish in total and tag them. The table shows their results when they return:

Name	Fish caught	Fish tagged
Alan	10	2
Bob	20	5
Caz	30	9

   What is the most accurate estimate for the population of fish? Explain your reasoning.

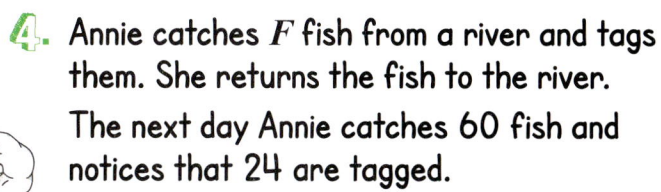

4. Annie catches F fish from a river and tags them. She returns the fish to the river.

   The next day Annie catches 60 fish and notices that 24 are tagged.

   Annie uses these results to estimate that there are 320 fish in the river.

   Work out the value of F.

CIRCLE THEOREMS 1

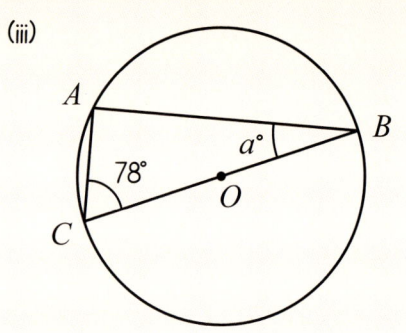

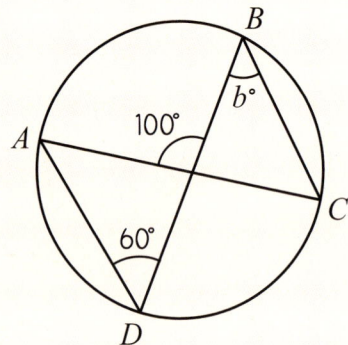

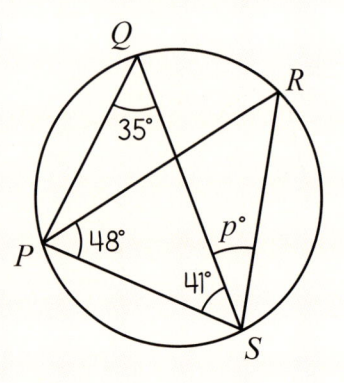

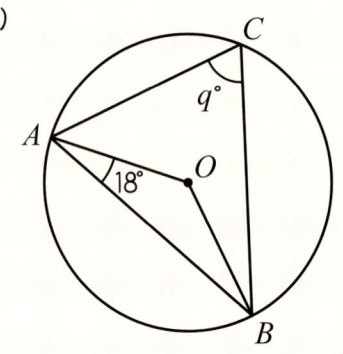

A.

(i) Circle with points S, R, P, Q. Angle at S = 57°, angle x° at R.

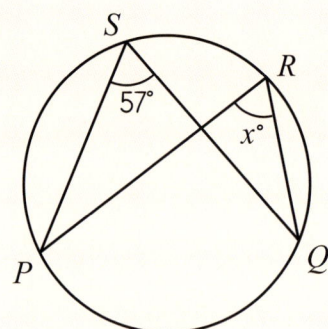

(ii) Circle with centre O, points M, N, L. Angle y° at N, angle 132° at O.

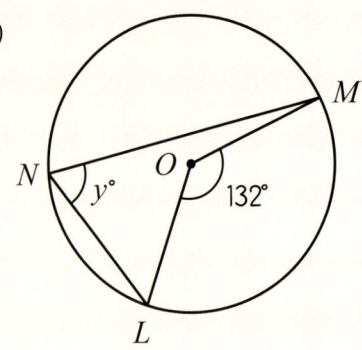

(iii) Circle with centre O, points A, B, C. Angle 78° at C, angle a° at B.

(iv) Circle with points A, B, C, D. Angle 100° at B (between BA and BC shown at interior), angle b° at B, angle 60° at D.

(v) Circle with points Q, R, P, S. Angle 35° at Q, angle 48° at P, angle 41° at S, angle p° at R.

(vi) Circle with centre O, points A, B, C. Angle q° at C, angle 18° at A.

B. Circle with points K, L, M, N, J. Angle 21° at K, angle 32° at J.

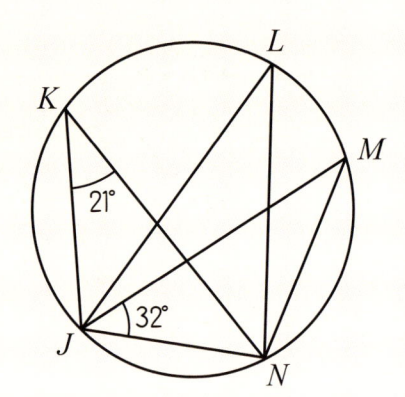

1. Circle with points Q, P, R, S. Angle x at Q, angle y at S.

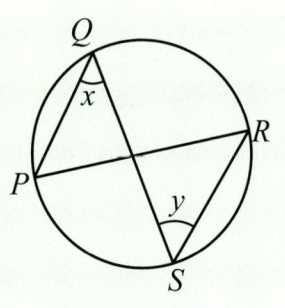

2. Circle with centre O, points A, B, C. Angle 76° at O, angle x° at O (below).

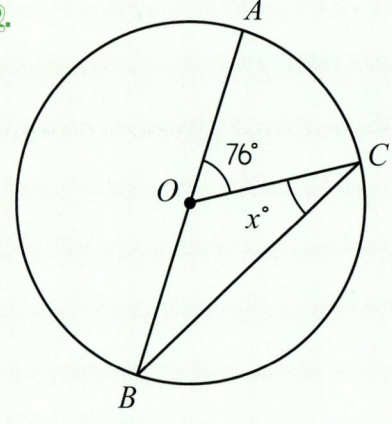

3. Circle with centre O, points L, N, M, P. Reflex angle 286° at O, angle 81° at P, angle a at M.

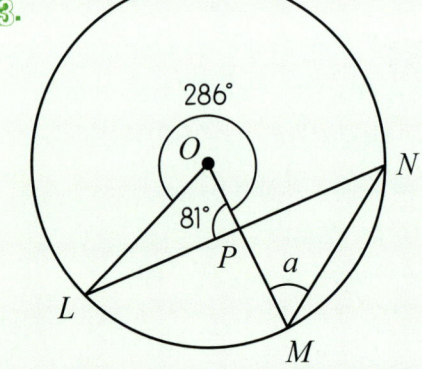

4. Circle with centre O, points A, B, C.

132

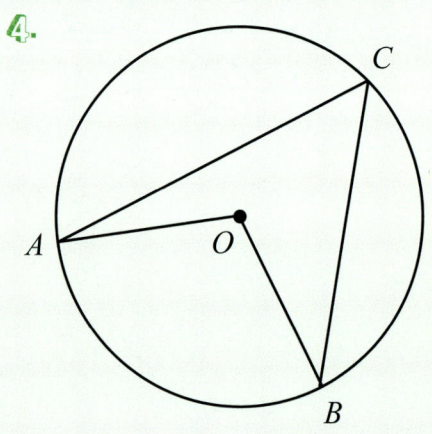

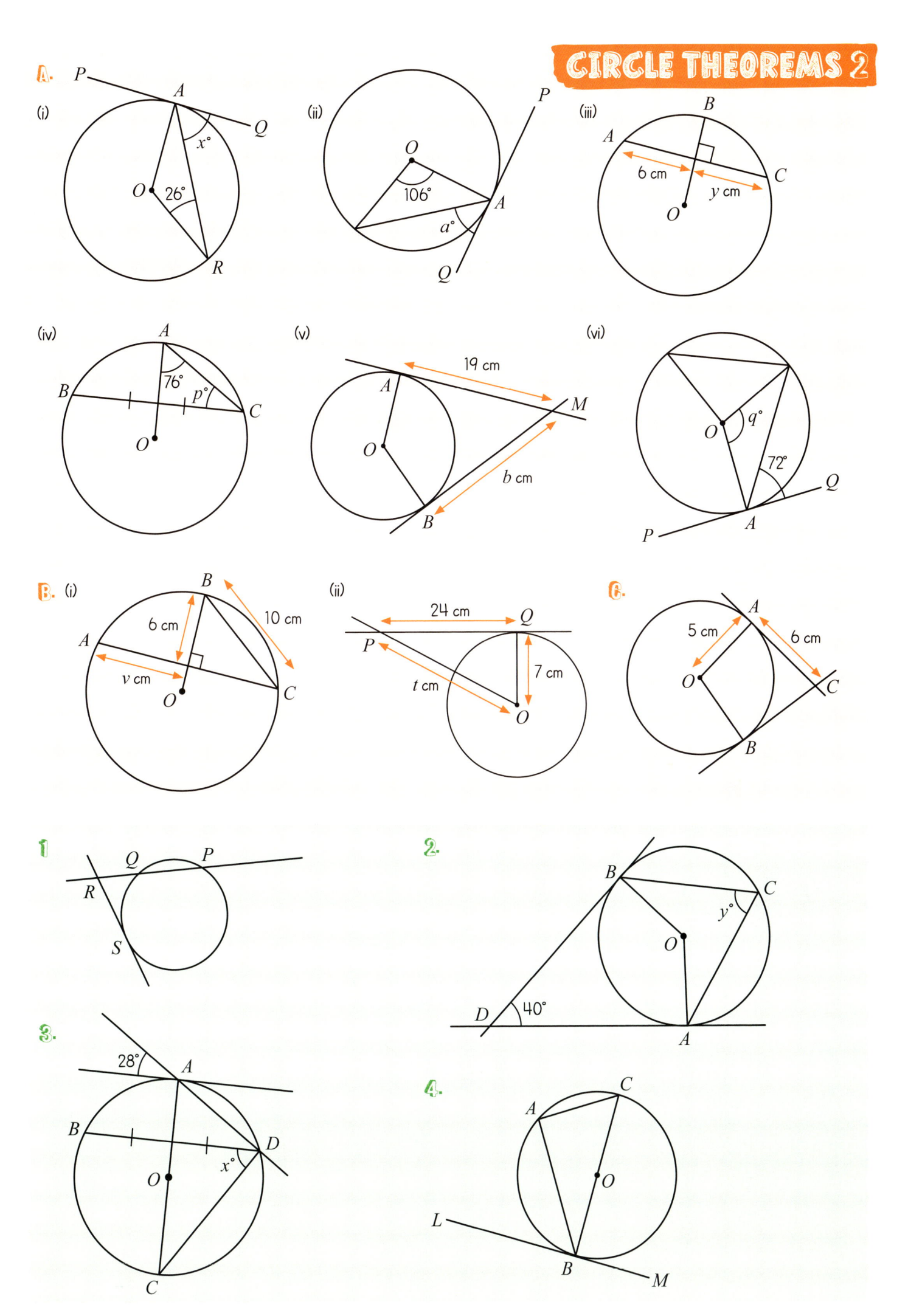

CIRCLE THEOREMS 2

A.

(i) Circle with centre O, tangent PAQ at A, angle ROA = 26°, angle x° between OA and AQ.

(ii) Circle with centre O, tangent PAQ at A, central angle 106°, angle a° at A.

(iii) Circle centre O, chord AC = 6 cm on one side of perpendicular from O, y cm on other side, B on circle.

(iv) Circle centre O, triangle ABC with angle A = 76°, AB = AC (tick marks), angle p° at C.

(v) Two tangents from M to circle centre O, touching at A and B. AM = 19 cm, BM = b cm.

(vi) Circle centre O, tangent PAQ at A, angle q° at O, angle 72° between chord and tangent at A.

B.

(i) Circle centre O, chord AC with perpendicular from B meeting at 6 cm from B, Av = v cm, BC = 10 cm.

(ii) Tangent from P touching circle at Q, PQ = 24 cm, OQ = 7 cm, PO = t cm.

C. Circle centre O, two tangents from external point meeting circle at A and B, OA = 5 cm, tangent length 6 cm to C.

1. Circle with two tangents/secants from external points through R, Q, P, S.

2. Circle centre O, tangent at A, line from D through A making 40° with tangent, chord to B and C, angle y° at C.

3. Circle centre O with diameter, tangent at A making 28° with chord, BD chord with tick marks BA = AD, angle x° at D.

4. Circle centre O, tangent LBM at B, chord to A and C, diameter through B.

CIRCLE THEOREMS 3

A.

(i) Circle with cyclic quadrilateral ABCD: angle A = 104°, angle B = 89°, angle C = a°.

(ii) Circle with tangent DE at A, chord AB and AC; angle ACB = 61°, angle ABC = 93°, angle s° between tangent and chord at A.

(iii) Cyclic quadrilateral ABCD: angle A = x°, angle B = 73°, angle D = y°, angle C = 112°.

(iv) Circle with points Q, R, P and tangent LPM at P: angle w° at Q, 91° at R, v° and 68° at P.

(v) Circle with tangent CQ and line CP at C; points A, B, D on circle; angle B = 79°, angle b° at A, 85° shaded at C, 61° between chord and tangent.

(vi) Circle with tangent MPN at P; points S, Q, R on circle; angle 52°, h° shaded at P, 41° at R.

B.

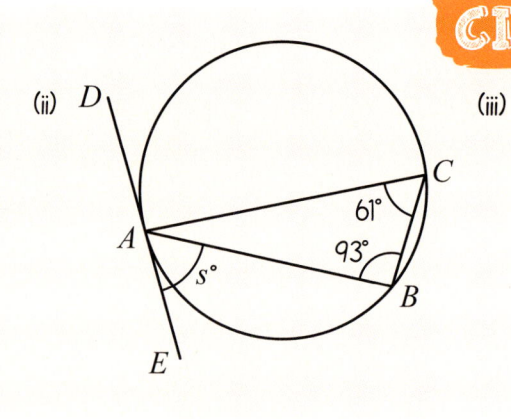

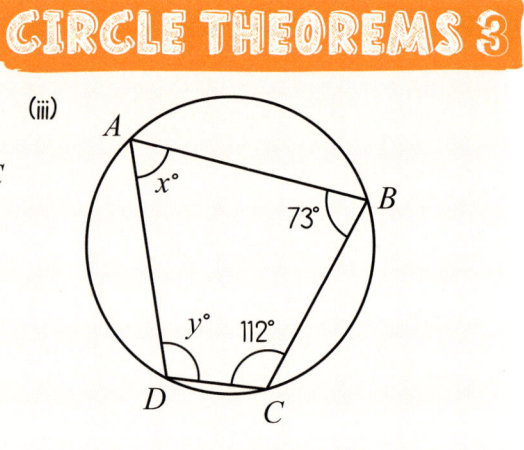

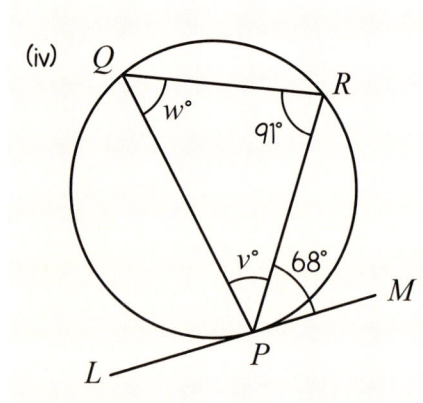

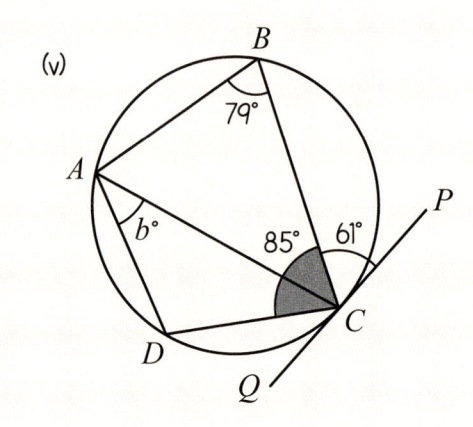

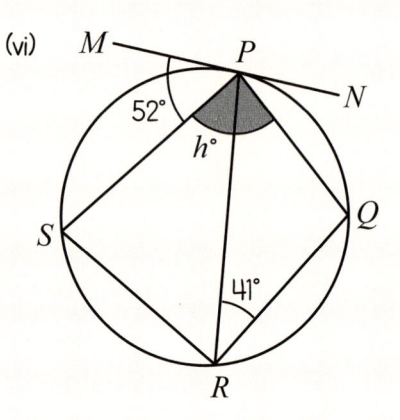

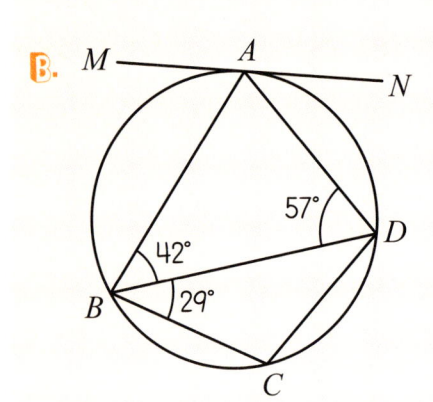

Circle with tangent MAN at A; points B, C, D on circle; angle ADB = 57°, angle ABD = 42°, angle DBC = 29°.

(i) Angle BAM

(ii) Angle BCD

(iii) Angle BDC

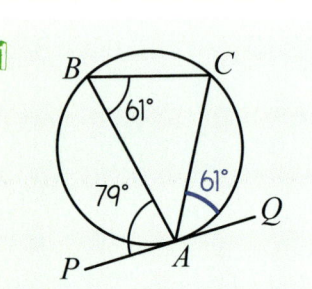

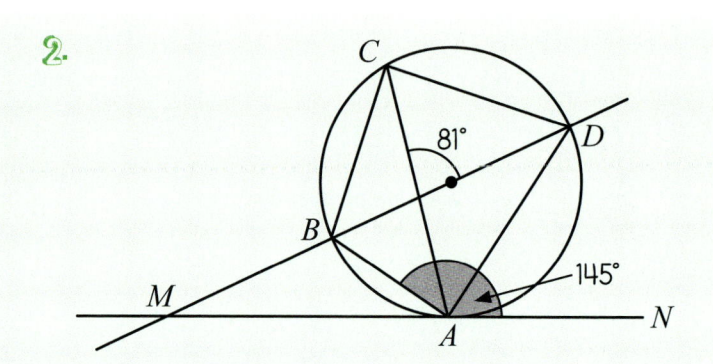

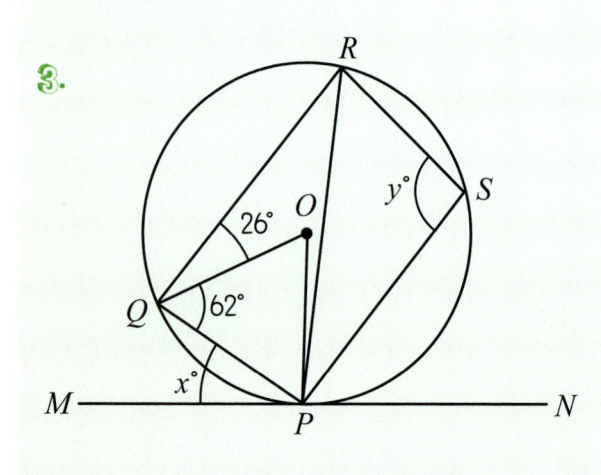

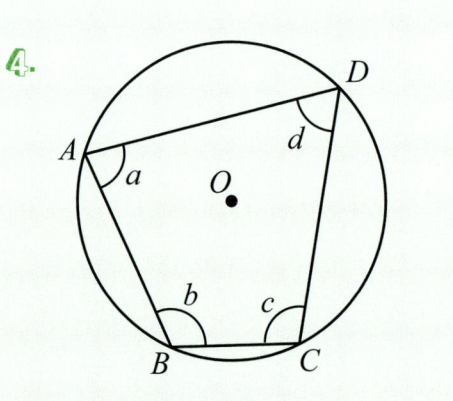

1. Circle with tangent PAQ at A; points B, C on circle; angle B = 61°, angle BAP = 79°, angle CAQ = 61°.

2. Circle with tangent MAN at A; points B, C, D on circle with BD through centre; angle C = 81°, angle 145° exterior at A.

3. Circle centre O; points Q, R, S, P on circle; angle OQR = 26°, angle OQP = 62°, angle y° at S, tangent MPN at P, angle x° at P.

4. Cyclic quadrilateral ABCD with centre O; angles a, b, c, d.

134

Solutions

Estimating with Powers & Roots (p.12)

Check-in

1	Rings around 8, 27, 64 and 1000
2	9 0.7 100 3
3	(100), 121, 144, 169 and 196

Set?

A	(i) 6.3, 6.4 or 6.5 (ii) 4.7, 4.8 or 4.9 (iii) 8.3, 8.4 or 8.5 (iv) 10.1, 10.2 or 10.3 (v) 2.6, 2.7 or 2.8 (vi) 4.1
B	216 000
C	40 000
D	91 125
E	1 336 336
F	(i) < (ii) < (iii) > (iv) <

Go!

1	Madge has divided by 3 and not taken the cube root. The cube root of 0.066 must be greater than 0.066 and less than 1
2	$50^3 = 125\,000$ (m^3) which is an underestimate as 53.1 was rounded down or $55^3 = 166\,375$ (m^3) which is an overestimate as 53.1 was rounded up or $53^3 = 148\,877$ (m^3) which is an underestimate as 53.1 was rounded down.
3	Using $\sqrt{752}$ leads to a side length of 27 or 28 metres and would give a perimeter of 108 or 112 metres respectively, so 108 or 112 metres are valid estimates.
4	For example: 845 watts $13^2 = 169$, so $130^2 = 16\,900$ and $16\,900 \div 20 = 845$ or 850 watts $13^2 = 169$, so $130^2 = 16\,900$ which could round to $17\,000$ and $17\,000 \div 20 = 850$

Recurring Decimals (p.14)

Check-in

1	$0.\dot{5}$ $0.4\dot{3}$ $0.0\dot{1}\dot{7}$ $0.\dot{7}5\dot{4}$
2	$\frac{4}{15}$ $\frac{13}{111}$ $\frac{68}{165}$

Set?

A	Ticks against: $\frac{5}{9}$ $\frac{9}{14}$ $\frac{4}{65}$
B	(i) $0.\dot{2}\dot{7}$ (ii) $0.58\dot{3}$ (iii) $0.2\dot{7}$ (iv) $0.7\dot{6}$
C	(i) $\frac{38}{99}$ (ii) $\frac{19}{33}$ (iii) $\frac{4}{15}$
D	(i) Let $x = 0.417417...$ So $1000x = 417.417...$ $x = \frac{417}{999}$ and $\frac{417}{999} = \frac{139}{333}$ (ii) Let $x = 0.92727...$ so $10x = 9.2727...$ and $1000x = 927.27...$ $x = \frac{918}{990}$ and $\frac{918}{990} = \frac{51}{55}$ (iii) Let $x = 0.346666...$ so $100x = 34.6666...$ and $1000x = 346.666...$ $x = \frac{312}{900}$ and $\frac{312}{900} = \frac{26}{75}$
E	(i) $\frac{13}{9}$ or $1\frac{4}{9}$ (ii) $\frac{98}{33}$ or $2\frac{32}{33}$ (iii) $\frac{1003}{330}$ or $3\frac{13}{330}$

Go!

1	No. Imogen has worked out $0.5\dot{4}$ The correct answer is $\frac{49}{90}$
2	170 runs = 99 balls $\frac{170}{99} = 1$ ball so $\frac{17000}{99} = 100$ balls $= 171.\dot{7}\dot{1}$
3	a) $a = 18$ b) $b = 211$ c) $c = \frac{257}{99}$ or $2.\dot{5}\dot{9}$ or $2\frac{59}{99}$
4	$1\frac{31}{270}$

Fractional Indices (p.16)

Check-in

1	9^7 9 9^{15} 9
2	0.125 or $\frac{1}{8}$
3	$9x^{10}y^4$ $3a^3b^5$

Set?

A	(i) 9 (ii) 3 (iii) 4 (iv) 3 (v) $\frac{1}{10}$ (vi) $\frac{2}{3}$
B	(i) 16 (ii) 8 (iii) 4 (iv) 125 (v) $\frac{25}{4}$ (vi) $\frac{81}{10\,000}$
C	(i) $\frac{1}{25}$ (ii) $\frac{1}{32}$ (iii) $\frac{1}{1000}$ (iv) $\frac{125}{27}$

Go!

1	Although the correct answer is 2, Nick has multiplied by one half and not found the square root of 4.
2	$\left(\frac{9}{4}\right)^{-\frac{3}{2}} = \frac{8}{27}$
3	a) $a = 81$ b) $b = 2$ c) $c = \frac{5}{4}$ or 1.25
4	$\sqrt[3]{4} \times \sqrt{8} = 2^{\frac{13}{6}}$

Product Rule for Counting (p.18)

Check-in

1	1600	2	380
3	1296	4	4154
5	512		
6	1320		

Set?

A	5850
B	110
C	1000
D	21
E	132 600

Go!

1	67 (girls)
2	He is not correct. He can reuse the letters and digits. The correct answer is: $26 \times 10 \times 10 \times 10 \times 26 = 676\,000$
3	$434 \times 24 \times 24 \times 24$ 5 999 616
4	264

135

Calculating with Bounds (p.20)

Check-in

1	18
2	1.26×10^2
3	2.475
4	50

Set?

A	42.5 (cm)
B	$\frac{1}{2}(6.45 + 12.25) \times 5.25$
	49.0875 (cm²)
C	$192\,500 \div 2150 = 89.534...$ so 90 people/km²

Go!

1	He should have worked out the lower bound of b subtract the upper bound of a.
	The correct working should be: $4.275 - 2.35 = 1.925$
2	a) -98
	b) 0.47
	c) 0.51
3	$1.55 \times 10^{11} \div 2.995 \times 10^8 = 517.529...$ so 518 seconds
4	2.4 (g/cm³) because both 2.427935... (UB) and 2.387091... (LB) round to 2.4 when rounded to 1 decimal place (or 2 significant figures).

Surds 1 (p.22)

Check-in

1	1 4 9 16 25 36 49 64 81 100 121 144
2	1, 4, 9 and 36

Set?

A	(i) $5\sqrt{2}$
	(ii) $6\sqrt{2}$
	(iii) $4\sqrt{2}$
	(iv) $5\sqrt{5}$
	(v) $6\sqrt{5}$
	(vi) $6\sqrt{3}$
B	(i) $\frac{4\sqrt{3}}{3}$
	(ii) $\frac{8\sqrt{5}}{5}$
C	(i) $\frac{9\sqrt{7}}{35}$
	(ii) $\frac{2\sqrt{2}}{3}$
	(iii) $\frac{4\sqrt{6}}{5}$
D	(i) $\frac{\sqrt{7}}{5}$
	(ii) $\frac{\sqrt{13}}{12}$
	(iii) $\frac{3\sqrt{3}}{8}$
	(iv) $\frac{2\sqrt{5}}{9}$

Go!

1	No. She has not used the largest factor of 48 that is a square number (16) so it is not simplified fully.
	The correct answer is $4\sqrt{3}$
2	$\sqrt{507} = \sqrt{3 \times 169}$
	$= \sqrt{3} \times \sqrt{169}$
	$= 13\sqrt{3}$
	So $k = 13$
3	$2\sqrt{33}$
4	$\sqrt{\frac{28}{49}} = \frac{\sqrt{28}}{\sqrt{49}}$
	$\frac{\sqrt{4} \times \sqrt{7}}{7} = \frac{2\sqrt{7}}{7}$
5	$\sqrt{54} = 3\sqrt{6}$

Surds 2 (p.24)

Check-in

1	$2\sqrt{2}$ $3\sqrt{6}$ $12\sqrt{2}$
2	$\frac{6\sqrt{5}}{5}$ $\frac{7\sqrt{3}}{6}$ $\frac{4\sqrt{2}}{9}$

Set?

A	(i) $4\sqrt{3} + \sqrt{21}$
	(ii) $9\sqrt{2} - 2\sqrt{15}$
	(iii) $24\sqrt{3} + \sqrt{6}$
	(iv) $\sqrt{70} - 5\sqrt{2}$
	(v) $3\sqrt{5} + 3\sqrt{13}$
B	(i) $18 + 7\sqrt{6}$
	(ii) $35 + 5\sqrt{3} - 7\sqrt{6} - 3\sqrt{2}$
	(iii) $10\sqrt{2} - 10 - 2\sqrt{7} + \sqrt{14}$
	(iv) $102 + 28\sqrt{3}$
	(v) $27 + 10\sqrt{2}$
C	(i) $\frac{18 + 7\sqrt{6}}{10}$
	(ii) $\frac{35 - 7\sqrt{6} + 5\sqrt{3} - 3\sqrt{2}}{19}$
	(iii) $\frac{102 + 28\sqrt{3}}{61}$

Go!

1	The correct answer is: $9a^2 - 6a\sqrt{b} + b$
2	$p = 3$
	$q = 7$
	$r = 5$
3	$b = 14$
	$c = 44$
4	a) $\left(\frac{1+\sqrt{5}}{2}\right)\left(\frac{1+\sqrt{5}}{2}\right) = \left(\frac{1}{2}+\frac{\sqrt{5}}{2}\right)\left(\frac{1}{2}+\frac{\sqrt{5}}{2}\right)$
	$= \frac{1}{4} + \frac{\sqrt{5}}{4} + \frac{\sqrt{5}}{4} + \frac{\sqrt{5}\sqrt{5}}{4} = \frac{1}{4} + \frac{5}{4} + \frac{2\sqrt{5}}{4}$
	$= \frac{6}{4} + \frac{\sqrt{5}}{2} = \frac{3}{2} + \frac{\sqrt{5}}{2}$
	and $\frac{1+\sqrt{5}}{2} + 1 = \frac{1}{2} + \frac{\sqrt{5}}{2} + \frac{2}{2}$
	$= \frac{3}{2} + \frac{\sqrt{5}}{2}$
	b) $1 \div \left(\frac{1+\sqrt{5}}{2}\right) = 1 \times \left(\frac{2}{1+\sqrt{5}}\right)$
	$= \frac{2}{1+\sqrt{5}}$
	Rationalising the denominator gives:
	$\left(\frac{2}{1+\sqrt{5}}\right)\left(\frac{1-\sqrt{5}}{1-\sqrt{5}}\right) = \frac{2-2\sqrt{5}}{1-5}$
	$= \frac{2-2\sqrt{5}}{-4} = \frac{1-\sqrt{5}}{-2} = \frac{\sqrt{5}-1}{2}$
	and $\frac{1+\sqrt{5}}{2} - 1 = \frac{1}{2} + \frac{\sqrt{5}}{2} - \frac{2}{2}$
	$= \frac{\sqrt{5}}{2} - \frac{1}{2} = \frac{\sqrt{5}-1}{2}$

Algebraic Fractions 1 (p.26)

Check-in

1	$\frac{23}{56}$	$\frac{13}{36}$	or equivalents
2	$8x + 18$		
	$3x - 59$		

Set?

A	(i) $\frac{11x + 23}{30}$	
	(ii) $\frac{16n - 15}{21}$	
	(iii) $\frac{8a - 11}{10}$	or equivalents
B	$\frac{x+6}{12}$ or equivalent	
C	(i) $\frac{4x + 14}{(x+4)(x+2)}$	
	(ii) $\frac{5y - 15}{y(y+5)}$	
	(iii) $\frac{p + 36}{(2p-3)(p+6)}$	
	(iv) $\frac{7x^2 - 23x}{(x-4)(x-3)}$	or equivalents
D	$\frac{3(x+3) + 4(3x-1)}{12} = \frac{3x + 9 + 12x - 4}{12} = \frac{15x + 5}{12} = \frac{5(3x+1)}{12}$	
	So $a = 5$, $b = 3$, $c = 1$ and $d = 12$	

Go!

1	$\frac{9n - 5m}{n(m - n)}$ or equivalent
2	$\frac{2x}{x+4} + \frac{x}{x+5} = \frac{x(3x+14)}{(x+4)(x+5)}$
3	He has made a mistake expanding the second bracket in his second step. The correct step should be: $\frac{9a + 6 - 8a + 2}{6}$ The correct solution should be $\frac{a+8}{6}$
4	$\frac{-13x + 3}{30}$ or equivalent

Solving Harder Equations (p.28)

Check-in

1	$x = -5$	2	$y = 2$
3	$m = 3$	4	$k = 8$

Set?

A	$x = \frac{1}{2}$ or 0.5
B	$y = \frac{1}{4}$ or 0.25
C	$m = -\frac{7}{8}$ or -0.875
D	$x = 4$
E	$w = \frac{2}{3}$
F	$a = -\frac{3}{4}$ or -0.75
G	$p = 3$
H	$n = \frac{1}{2}$ or 0.5
I	$v = -\frac{4}{25}$

Go!

1	$n = \frac{1}{30}$
2	a) $a = 6$ b) $b = \frac{6}{7}$
3	a) $x = \frac{5}{4}$ or $1\frac{1}{4}$ or 1.25 b) 7.5 cm
4	$(0.5, 6)$

Factorising (p.30)

Check-in

1	$(x + 3)(x + 4)$ $(y - 4)(y - 5)$ $(a + 3)(a - 6)$ or equivalents
2	$(p - 9)(p + 9)$ or equivalent

Set?

A	(i) $(2x + 7)(x + 2)$ (ii) $(2y + 9)(y + 3)$ (iii) $(3x + 4)(x + 1)$ (iv) $(11p + 2)(p + 4)$ or equivalents
B	(i) $(3x - 2)(x + 5)$ (ii) $(2n + 3)(n - 4)$ (iii) $(5w - 3)(w - 2)$ (iv) $(7x + 1)(x - 5)$ or equivalents
C	(i) $(5x - 6)(5x + 6)$ (ii) $(12c - 11)(12c + 11)$ or equivalents

Go!

1	This is not always true. If the coefficient of y^2 was 1 then it would be true. $3y^2 + 19y + 28$ can be factorised to $(3y + 7)(y + 4)$
2	$(3a^2 - 2b)(3a^2 + 2b)$ or equivalent
3	$2(x - 3)(3x - 4)$ or equivalent
4	$(-2x - 1)(x - 3)$ or $(3 - x)(2x + 1)$ or $-(x - 3)(2x + 1)$ or equivalent
5	$2x^2 - 5x - 12 = (2x + 3)(x - 4)$

Expanding Brackets (p.32)

Check-in

1	$x^2 + 11x + 28$ or equivalent
2	$x^2 - 5x - 24$ or equivalent
3	$y^2 - 7y + 10$ or equivalent
4	$2x^2 - 11x - 6$ or equivalent
5	$6p^2 + 39p + 63$ or equivalent

Set?

A	(i) $x^3 + 10x^2 + 29x + 20$ (ii) $y^3 + 8y^2 + y - 42$ (iii) $x^3 - 5x^2 - 34x + 80$ or equivalents
B	(i) $2x^3 + 23x^2 + 59x + 24$ (ii) $5p^3 + 53p^2 + 118p - 56$ (iii) $21n^3 - 125n^2 + 182n - 72$ or equivalents
C	For example: $(10x^2 - 31x - 14)(3x + 8)$ $30x^3 + 80x^2 - 93x^2 - 248x - 42x - 112$ $30x^3 - 13x^2 - 290x - 112$ $a = 30$, $b = -13$, $c = -290$ and $d = -112$

Go!

1	$2x^3 - 15x^2 + 13x + 60$ $(2x + 3)(x - 4)(x - 5)$ or $(2x + 3)(x - 5)(x - 4)$
2	$9n^3 - 6n^2 - 32n + 32$
3	$81k^4 - 16$
4	$8x^3 + 12x^2y + 6xy^2 + y^3$
5	For example: $(x + 3)(x - 6)(2x + 13) = 1716$ $(x^2 - 3x - 18)(2x + 13) = 1716$ $2x^3 + 13x^2 - 6x^2 - 39x - 36x - 234 = 1716$ $2x^3 + 7x^2 - 75x = 1950$

Algebraic Fractions 2 (p.34)

Check-in

1	$\frac{7x - 24}{10}$ $\frac{3x^2 + x}{(x+3)(x-1)}$
2	$(3x + 4)(x + 1)$ $(2x - 9)(2x + 9)$ or equivalents

Set?

A	(i) $\frac{3}{x - 2}$ (ii) $\frac{4n}{n - 6}$ (iii) $\frac{3x - 4}{x + 6}$ (iv) $\frac{x + 6}{3x - 5}$
B	(i) $\frac{7x + 4}{x + 7}$ so $a = 7$, $b = 4$, $c = 7$ (ii) $\frac{4x - 1}{x - 2}$ so $a = 4$, $b = -1$, $c = -2$
C	$\frac{18k}{k - 5}$

Go!

1	y is not a factor of all four terms in the second step. The final answer should be: $\frac{y + 9}{2y + 3}$
2	$\frac{-1}{(x + 6)}$
3	$\frac{11x + 7}{2x + 1}$ $a = 11$, $b = 7$, $c = 2$ and $d = 1$
4	$\frac{4w + 7}{w - 6}$

137

Algebraic Proof (p.36)

Check-in

1	$20n - 16$
2	$-4p + 38$
3	$2m^2 + 7m + 3$
4	$4p^2 + 8p + 3$
5	$4a^2 + 4a + 1$

Set?

A	For example: (i) $2n + 1, 2n + 3$ (ii) n^2 (iii) $n^2, (n + 1)^2$ (iv) $(2n + 1)^2$ (v) $\frac{1}{2}n(n + 1), \frac{1}{2}m(m + 1)$
B	For example: $n + n + 1 + n + 2 + n + 3$ $= 4n + 6$ $= 2(2n + 3)$ Even numbers are multiples of 2
C	For example: $4p^2 + 4 + 4p^2 + 16p + 16$ $= 8p^2 + 16p + 20$ $= 4(2p^2 + 4p + 5)$ which is a multiple of 4
D	For example: $15x - 10 + 18x - 12$ $= 33x - 22$ $= 11(3x - 2)$ which is a multiple of 11
E	For example: $16n^2 + 24n + 9 + 9n^2 + 24n + 16 - n^2$ $= 24n^2 + 48n + 25$ $= 2(12n^2 + 24n + 12) + 1$ which means two lots of "something" plus 1 will always be odd, or $= 24n^2 + 48n + 25$ $=$ even + even + odd which will always be odd.
F	even × odd × even = even or odd × even × odd = even

Go!

1	For example: $(6y + 5)^2 - (6y - 5)^2$ $= 36y^2 + 60y + 25 - (36y^2 - 60y + 25)$ $= 120y$ $= 24(5y)$ which is a multiple of 24
2	For example: $2n + 1$ is always odd So $(2n + 1)(2n + 1)(2n + 1)$ $= (4n^2 + 4n + 1)(2n + 1)$ $= 8n^3 + 12n^2 + 6n + 1$ $= 2(4n^3 + 6n^2 + 3n) + 1$ which means two lots of "something" plus 1 will always be odd.
3	For example: The nth term rule is $4n - 1$ so any term is $4n - 1$ and any subsequent term will be $(4n - 1) + 4$ The difference between the squares will be: $(4n - 1 + 4)^2 - (4n - 1)^2$ $= (4n + 3)^2 - (4n - 1)^2$ $= 16n^2 + 24n + 9 - 16n^2 + 8n - 1$ $= 32n + 8 = 8(4n + 1)$ which is a multiple of 8
4	$\frac{1}{2}(k^2 + 3k + 2) + \frac{1}{2}(k^2 + 5k + 6)$ $= \frac{1}{2}(2k^2 + 8k + 8)$ $= \frac{1}{2} \times 2(k^2 + 4k + 4)$ $= (k^2 + 4k + 4)$ $= (k + 2)^2$ which is a square number.

Functions 1 (p.38)

Check-in

1	145	3
2	$x = 6.2$	$x = 15$

Set?

A	(i) 13	(ii) 34	
	(iii) -2	(iv) -14	
B	(i) 16	(ii) 71	
	(iii) 8	(iv) $7\frac{9}{16}$ or $\frac{121}{16}$	
C	(i) 9	(ii) 6	
D	$x = 15$		
E	$x = 3.5$		
F	$x = 8$		
G	(i) $x = 9$		
	(ii) $x = 10$		
	(iii) $x = 7.6$		

Go!

1	$f(x) = 3x^2 + 4$ $f(x) = 79$ when $x = 5$
2	$k = 4$
3	A function should have one output for every possible input. When $x = 0$, there is no output.
4	a) $8a^2 - 3$ b) $2a^2 + 8a + 5$ c) $8a^2 + 16a + 5$

Functions 2 (p.40)

Check-in

1	$x = \frac{y}{6}$	$x = y + 45$	
	$x = \frac{y - 4}{7}$	$x = \frac{8y + 10}{5}$	or equivalents

Set?

A	(i) $f^{-1}(x) = \frac{x}{5}$ (ii) $f^{-1}(x) = x - 6$ (iii) $f^{-1}(x) = \frac{x + 17}{2}$ (iv) $f^{-1}(x) = \frac{x - 2}{7}$ (v) $f^{-1}(x) = \frac{x + 6}{3}$ or $f^{-1}(x) = \frac{x}{3} + 2$
B	(i) $g^{-1}(x) = 7x$ (ii) $g^{-1}(x) = 8x - 13$ (iii) $g^{-1}(x) = 15x + 4$ (iv) $g^{-1}(x) = \frac{2x - 7}{8}$
C	(i) $fg(x) = 4x^2 + 1$ (ii) $hf(x) = \frac{12x - 8}{5}$ (iii) $gf(x) = 16x^2 - 24x + 10$ (iv) $hg(x) = \frac{3x^2 + 4}{5}$ (v) $ff(x) = 16x - 15$

Go!

1	Yes, she has used y instead of $f(x)$ in her working which is valid.
2	$f^{-1}(x) = \frac{6x - 5}{-2}$ or $f^{-1}(x) = \frac{5 - 6x}{2}$
3	$f^{-1}(x) = \frac{3x + 1}{5}$ $ff^{-1}(x) = \frac{5\left(\frac{3x + 1}{5}\right) - 1}{3}$ $= \frac{\left(\frac{15x + 5}{5}\right) - 1}{3}$ $= \frac{3x + 1 - 1}{3} = x$
4	$f^{-1}(x) = \frac{x - 6}{6}$ $g^{-1}(x) = \frac{2x - 7}{2}$ $\frac{x - 6}{6} = \frac{2x - 7}{2}$ can be solved to give $x = 3$
5	$fg(x) = 9x^2 + 36x + 36$ $gf(x) = 3x^2 + 12x + 16$ $3gf(x) = 9x^2 + 36x + 48$ $9x^2 + 36x + 48 - (9x^2 + 36x + 36) = 12$

Perpendicular Lines (p.42)

Check-in

1	$\frac{1}{4}$	7	$-\frac{9}{4}$	$\frac{5}{2}$ or 2.5
2	$x = \frac{y-2}{7}$		$x = 2y - 14$ or $x = 2(y-7)$	
3	gradient = 5		gradient = $-\frac{1}{3}$	
	y-intercept = 8		y-intercept = 12	

Set?

A
(i) $-\frac{1}{6}$
(ii) $\frac{1}{15}$
(iii) 5
(iv) -8
(v) $-\frac{3}{2}$
(vi) $\frac{9}{4}$

B Ticks against:
$y = 6x - \frac{1}{6}$ and $y = -\frac{1}{6}x - 7$

C
$y = -\frac{1}{3}x + 4$ matched with $y = 3x - \frac{1}{3}$
$y = 3 - 4x$ matched with $y = \frac{1}{4}x + 3$
$y = -\frac{1}{4}x + 3$ matched with $y = 4x + 3$

D $y = -3x - 8$

Go!

1	Rearranging the second equation to be in the form $y = mx + c$ gives a gradient of $-\frac{5}{2}$ which is not the negative reciprocal of $\frac{1}{5}$
2	$x + 4y = 16$ and $4x - y + 9 = 0$
3	Gradient of line A = $\frac{4}{3}$
	Gradient of line B = $-\frac{3}{4}$
	The equation of line B is $y = -\frac{3}{4}x + c$
	Substituting in $(8, -2)$ we find $c = 4$ so $y = -\frac{3}{4}x + 4$
4	$y = -\frac{3}{5}x + \frac{53}{5}$ or equivalent

Completing the Square (p.44)

Check-in

1	$(x + 4)^2$ or $(x + 4)(x + 4)$
2	$(x + 2)(x + 4)$ or equivalent
3	$9p(p + 4)$
4	$x^2 + 12x + 27$
5	$\frac{25}{4}$ or 6.25
6	$\frac{17}{4}$ or 4.25

Set?

A
(i) $(x + 7)^2 - 49$
(ii) $(x - 10)^2 - 100$
(iii) $\left(x + \frac{5}{2}\right)^2 - \frac{25}{4}$ or equivalent
(iv) $\left(x - \frac{9}{2}\right)^2 - \frac{81}{4}$ or equivalent

B
(i) $(x + 2)^2 + 8$
(ii) $(x - 3)^2 - 5$
(iii) $(x + 6)^2 - 38$
(iv) $(x - 4)^2 - 21$

C
(i) $\left(x + \frac{7}{2}\right)^2 - \frac{53}{4}$ or equivalent
(ii) $\left(x - \frac{5}{2}\right)^2 - \frac{13}{4}$ or equivalent

Go!

1	$x^2 + 6x + 8 = (x + 3)^2 - 1$ or
	$x^2 + 6x + 1 = (x + 3)^2 - 8$
2	$\left(x + \frac{7}{2}\right)^2 - \frac{9}{4}$ or equivalent
3	$(x - 9)^2 - 78 = 0$
	$(x - 9)^2 = 78$
	$x - 9 = \pm\sqrt{78}$
	$x = 9 \pm \sqrt{78}$
4	$2(x - 3)^2 - 13$

Quadratic Functions (p.46)

Check-in

1

x	-2	-1	0	1	2	3
y	7	0	-5	-8	-9	-8

2	$x = 1$ and $x = -7$
3	$(x + 5)^2 - 27$
4	Roots: $x = -1$ and $x = 3$
	Turning point: $(1, -1)$

Set?

A
(i) $(2, 13)$
(ii) $x = -0.6$ to -0.5 and $x = 4.5$ to 4.6

B
(i) $(x + 7)^2 - 54$
(ii) $(-7, -54)$

C $(4.5, 3.75)$ or equivalent

Go!

1	No. Completing the square gives $(x - 8)^2 - 67$ which results in a turning point of $(8, -67)$

2 a)

x	-1.5	-1	-0.5	0	0.5	1	1.5	2	2.5
y	-3.25	2	5.75	8	8.75	8	5.75	2	-3.25

b)

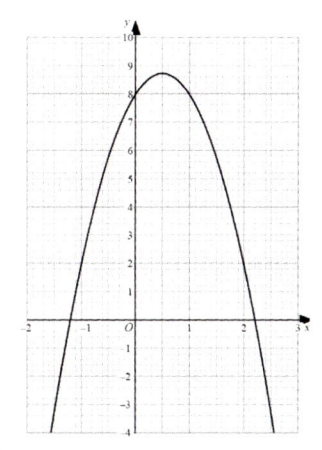

c) $(0.5, 8.75)$
d) $x = -1.3$ to -1.1 and $x = 2.1$ to 2.3
e) $x = -1.3$ to -1.2 and $x = 1.8$ to 1.95
(This is where $y = 8 + 3x - 3x^2$ and $y = x + 1$ intersect).

3
a) $\left(x - \frac{5}{2}\right)^2 - \frac{81}{4}$ or equivalent
b) $\left(\frac{5}{2}, -\frac{81}{4}\right)$ or equivalent
c) $x = -2$ and $x = 7$
d)

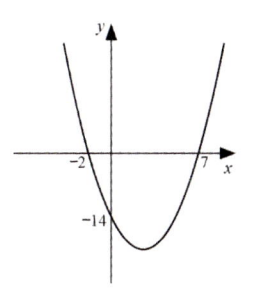

Quadratic Equations 1 (p.48)

Check-in

1	$x = 7$ and $x = 2$
2	$y = -6$ and $y = 0$
3	$x = -8$ and $x = 4$
4	$x = -12$ and $x = -3$

Set?

A	$x = \frac{2}{3}$ and $x = -\frac{2}{3}$
B	$p = \frac{10}{9}$ and $p = -\frac{10}{9}$
C	$x = -\frac{2}{3}$ and $x = 7$
D	$y = 3$ and $y = \frac{7}{5}$
E	$x = -8$ and $x = \frac{1}{2}$
F	$n = -10$ and $n = -\frac{2}{7}$
G	$x = 5$ and $x = \frac{1}{3}$
H	$k = 4$ and $k = \frac{3}{2}$

Go!

1	$x = \frac{6}{5}$ and $x = -\frac{6}{5}$
2	$a = -5$ and $a = \frac{4}{3}$
3	$y = -2$ and $y = -\frac{5}{2}$
4	No, the correct solutions are $x = \frac{3}{5}$ and $x = -4$
5	$7x^2 - 47x + 30 = 0$ $x = 6$ and $x = \frac{5}{7}$

Quadratic Equations 2 (p.50)

Check-in

1	$x = 7$ and $x = 8$
2	$x = -2$ and $x = \frac{4}{5}$
3	$\left(x + \frac{3}{2}\right)^2 - \frac{33}{4}$

Set?

A	Ticks against: $3k^2 - 11k + 9 = 0$ $5x^2 - 9x - 7 = 0$
B	(i) $a = 4, b = 7, c = 2$ (ii) $a = 4, b = -7, c = 2$ (iii) $a = 4, b = 7, c = -2$ (iv) $a = -4, b = -7, c = 2$
C	$x = -1.43$ and $x = -0.23$
D	$x = -4.04$ and $x = 0.21$
E	$x = 0.45$ and $x = 1.26$
F	$x = -0.55$ and $x = 0.68$
G	$x = 0.74$ and $x = 2.26$

Go!

1	$x = -1.45$ and $x - 1.95$
2	$x = -1.65$ and $x = 1.45$
3	Gabriel has placed the first negative symbol in the formula incorrectly leading to one of the solutions being incorrect. The correct answers should be $x = -1.64$ and $x = 0.24$
4	a) $x = 3.5$ Pythagoras' Theorem gives: $(4x - 7)^2 + (3(2x + 1))^2 = (4x + 11)^2$ Expanding, simplifying and solving gives two solutions (the answer must be positive and so $x = -0.5$ is disregarded). b) 25 (cm)
5	$x = -0.301$ and $x = 8.30$

Exponential Graphs (p.52)

Check-in

1	625	$\frac{1}{16}$	$\frac{1}{10000}$ or 0.0001
	81	$\frac{1}{81}$	$\frac{1}{10000}$ or 0.0001
	81.6293376	$\frac{25}{2}$ or 12.5	

Set?

A

x	-2	-1	0	1	2	3	4
y	0.25	0.5	1	2	4	8	16

x	-2	-1.5	-1	-0.5	0	0.5	1
y	25	11.2	5	2.2	1	0.4	0.2

Some answers given to 1 decimal place

x	-2	-1	0	1	2	3	4
y	0.625	1.25	2.5	5	10	20	40

Go!

1.

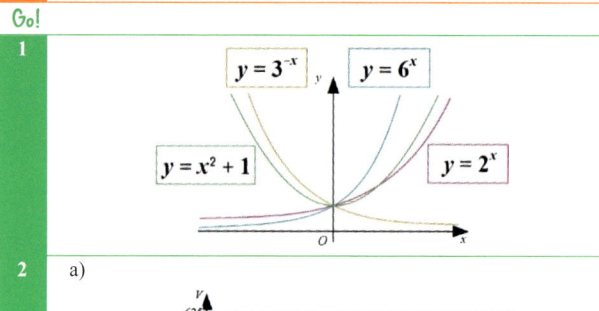

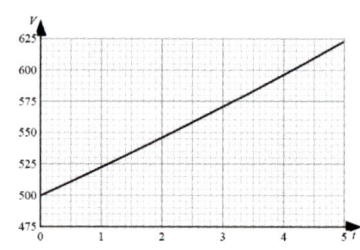

2. a)

b) The function is an exponential function because the variable is in the power (even through this small section of the graph may appear to be a straight line).

3. a) The initial mass of the radioactive substance when time is 0
   b)

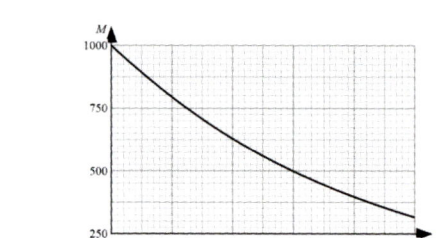

4. $A = 10$ and $B = 2$

Trigonometric Graphs (p.54)

Check-in

1	1	0.87
	0.71	0.5
	0	−0.5
	−0.87	−1

Set?

A
(i) 14 (accept 13 to 15) and 166 (accept 165 to 167)
(ii) 40.5 (accept 40 to 41) and 139.5 (accept 139 to 140)

B [cosine-like graph from 0 to 360]

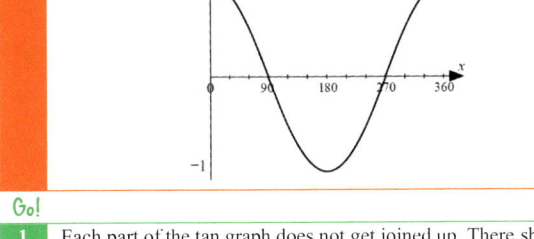

Go!

1 Each part of the tan graph does not get joined up. There should be asymptotes at 90° and 270°

2 [graph]

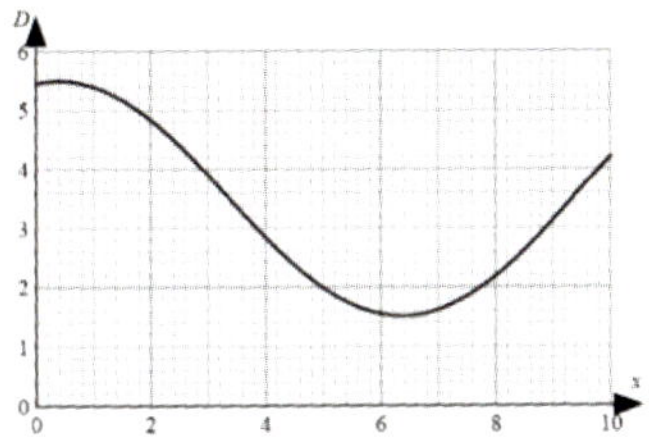

3
a) The graph of tan x repeats every 180°
tan 60° = tan 240° = tan 420° = tan 600°
b) The part of the sin graph between 0 and 180° has a line of symmetry at 90°
[graph with markings at 15, 165, 180]

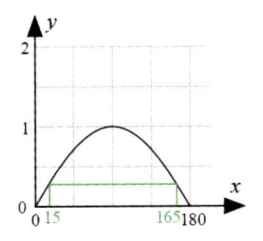

4 Tick to indicate 1 correct solution.
[graph]

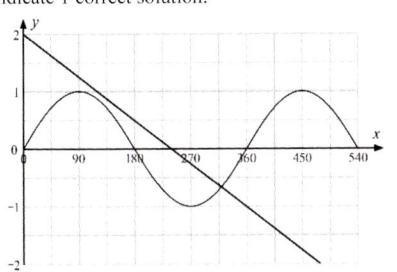

Transforming Graphs 1 (p.56)

Check-in

1 [graph showing triangle A translated]

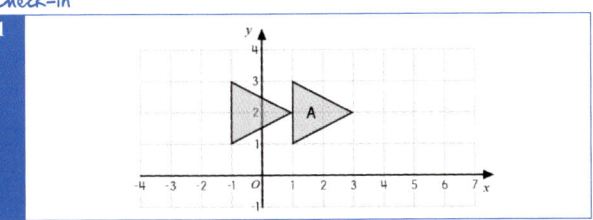

Set?

A
(i) $y = f(x + 4)$
(ii) $y = f(x) - 2$

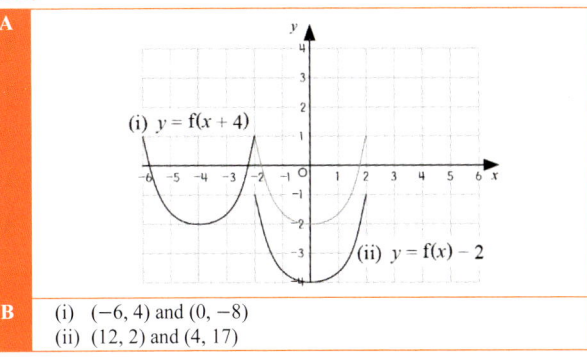

B
(i) (−6, 4) and (0, −8)
(ii) (12, 2) and (4, 17)

Go!

1 Bev should have translated the graph 4 units to the left.

2 [graph with (−1, 3) and (5, 2), $y = f(x)$]
Turning point (−1, 3)

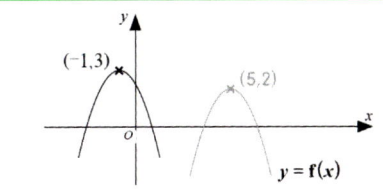

3 $x = -2$ (unchanged) and $y = -14$

4 Statements ticked:
sin x = cos (x − 90°) and cos x = sin (x + 90°)

Transforming Graphs 2 (p.58)

Check-in

1 [graph with rectangle A]

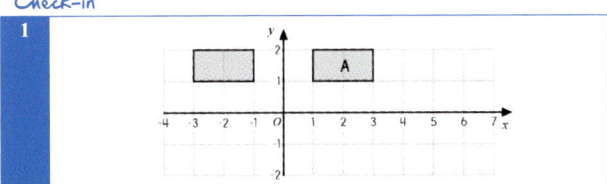

Set?

A
(i) $y = f(-x)$
(ii) $y = -f(x)$

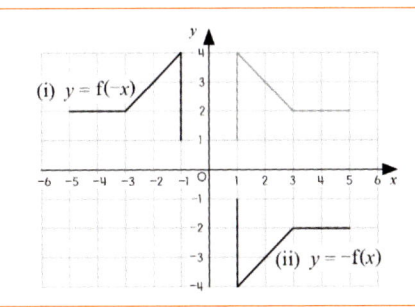

B

	A	B
$y = f(-x)$	(0, 2)	(−2, 2)
$y = -(x)$	(0, −2)	(2, −2)

Go!

1 [graph]

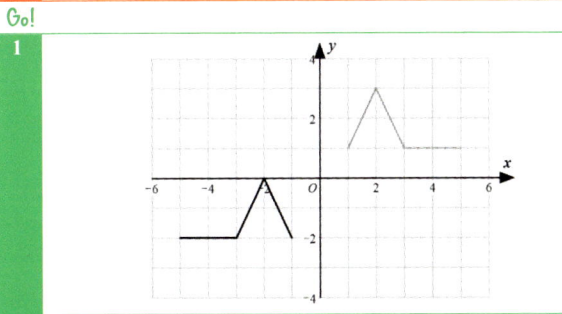

2
a) (2, −1)
b) (−2, 7)
c) (−1, −4)

3 $k = 4$

4 Odd
Even
Odd
Odd
Even

141

Estimating Gradients (p.60)

Check-in

1	4	$\frac{1}{4}$ or 0.25
	-2	$-\frac{1}{2}$ or -0.5
2	$A = -2$	
	$B = \frac{1}{2}$ or 0.5	
	$C = \frac{2}{3}$	

Set?

A	(i) Range accepted -2.9 to -2.5
	(ii) Range accepted 0.4 to 0.6
	(iii) Range accepted -2.9 to -2.5

Go!

1	The gradient of a curve is constantly changing so you cannot say the gradient of a curve is a fixed number.
2	There will be a positive gradient at (5, 1).
	There will be negative gradients at (0, 11) and (2, 1).
	(3.5, -1.25) is the turning point so the gradient is zero here.
	(5, 1) is the point at which the gradient will be 3
3	a) Range accepted -5 to -4.7
	b) A coordinate with an x value between 2 and 2.1 and a y value between 2.7 and 3.3

Rates of Change (p.62)

Check-in

1	Range accepted 1.2 to 1.5
	Range accepted -0.8 to -0.7

Set?

A	(i) Range accepted 0.7 to 0.9 m/s
	(ii) Approximately 1 m/s (accept approximately -1 m/s)
	(iii) Approximately 1 m/s (accept approximately -1 m/s)

Go!

1	a) Range accepted 32 to 35 (m/s^2)
	b) The acceleration at time $t = 1$ second.
2	October because it is the steepest part of the graph at this point.
3	a) Range accepted 40 to 41 km/h
	b) $140 \div 4.5 = 31.\dot{1}$ km/h
	c) Any time between 1.2 hours and 1.6 hours (steepest)
	d) Ranges accepted:
	0.3 to 0.5 hours and
	1.9 to 2 hours and
	2.8 to 3 hours.

Area under a Graph (p.64)

Check-in

1	50 cm^2	32.4 cm^2

Set?

A	(i) Range accepted 29.3 to 29.7 metres
	(ii) Underestimate because between 0 and 6 seconds the strips are lower than the curve.
B	43 metres

Go!

1	No, he should have written:
	Approximately 10,575 litres flows through the gate in the first 40 seconds.
2	a) 5 seconds
	b) Range accepted 5.2 to 6 m/s^2 (the gradient of the tangent to the curve at 2 seconds)
	c) Range accepted 165 to 167 metres
3	(velocity-time graph: rises from (0,0) to (1,10), flat to (1.5,10), rises to (2.5,15), flat to (3.5,15), falls to (4.5,0))

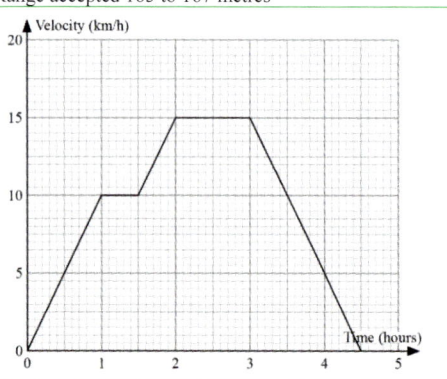

Simultaneous Equations 1 (p.66)

Check-in

1	$x = 5$ and $y = 4$	2	$x = -2$ and $y = 5$
3	$a = -1$ and $b = -3$	4	$p = 3$ and $q = 0.5$

Set?

A	$x = 12$ and $y = 7$
B	$x = 4$ and $y = -8$
C	$p = -6$ and $q = \frac{2}{5}$
D	$x = -4$ and $y = -16$
E	$a = -10$ and $b = 9$
F	$x = \frac{2}{3}$ and $y = -\frac{1}{5}$

Go!

1	$3.2 < x < 3.3$
	$0.8 < y < 0.9$
2	Nargis has not multiplied the fraction correctly.
	$\frac{2}{5} \times 5 = 2$ (not 10)
3	The second line of his working should be:
	$5x - 6x + 4 = -3$
	(which leads to $x = 7$ and $y = 19$)
4	$a = -3$ and $b = \frac{1}{3}$

Simultaneous Equations 2 (p.68)

Check-in

1	$x = 4$ and $y = 2$
2	$x = -4.5$ and $x = 1$
3	$x = -1.12$ and $x = 0.45$

Set?

A	$x = -2, y = 11$ and $x = 6, y = 3$
B	$x = -1, y = 5$ and $x = 5, y = 17$
C	$x = -2.29, y = 8.59$ and $x = 0.54, y = 2.91$
D	$x = 3.66, y = -1.68$ and $x = 5.65, y = 2.31$
E	$x = -1.68, y = 1.32$ and $x = 0.48, y = 3.48$

Go!

1	No because Dale has only found one of the two sets of solutions.
	The other solution is $x = -11$ and $y = 28$
2	$x = \frac{1}{2}, y = \frac{13}{2}$ and $x = 3, y = -1$
3	38.5 (cm)
4	$x = -\frac{56}{13}, y = -\frac{15}{13}$ and $x = 4, y = 3$

Rearranging Formulae (p.70)

Check-in

1	$x = \frac{y - 6}{2}$	$x = \frac{y - 12}{2}$ or $x = \frac{y}{2} - 6$
	$x = 2y - 6$	$x = 2(y - 6)$
	$x = \frac{2y - 6}{5}$	$x = \frac{2y - 30}{5}$ or $x = \frac{2y}{5} - 6$

Set?

A	(i) $x = \frac{1}{y - 3}$
	(ii) $x = \frac{-3}{y - 4}$
	(iii) $x = \frac{b}{y - a}$
	(iv) $x = \frac{7}{y + 4}$
	(v) $x = \frac{1}{9y - 6}$ or equivalents
B	(i) $a = \frac{12 + 7T}{T - 1}$
	(ii) $a = \frac{20 - 6b}{b - 1}$
	(iii) $a = \frac{-bx - db}{d - x}$
	(iv) $a = \frac{2A}{h} - b$ or $a = \frac{2A - hb}{h}$ or equivalents
C	$x^3 - 6x + 23 = 0$
	$x^3 + 23 = 6x$
	So $x = \frac{x^3 + 23}{6}$

D
$x^3 - 4x^2 - 7 = 0$
$x^3 = 7 + 4x^2$
Dividing both sides by x^2 gives $x = \frac{7 + 4x^2}{x^2}$
so $x = \frac{7}{x^2} + 4$

Go!

1 a) $x^3 = 2x + 9$
$x = \sqrt[3]{2x + 9}$
b) $2x = x^3 - 9$
$x = \frac{x^3 - 9}{2}$
c) $x(x^2 - 2) = 9$
$x = \frac{9}{x^2 - 2}$

2 $x^2 - 3 = \frac{4}{x}$
$x^3 - 3x = 4$
$x^3 = 3x + 4$
$x = \sqrt[3]{3x + 4}$

3 $a = \frac{s - ut}{\frac{1}{2}t^2} = \frac{2(s - ut)}{t^2}$

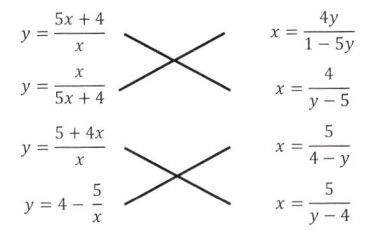

4
$y = \frac{5x + 4}{x}$ $x = \frac{4y}{1 - 5y}$
$y = \frac{x}{5x + 4}$ $x = \frac{4}{y - 5}$
$y = \frac{5 + 4x}{x}$ $x = \frac{5}{4 - y}$
$y = 4 - \frac{5}{x}$ $x = \frac{5}{y - 4}$

5 $p = \frac{rq}{3q - 2r}$ or equivalent

Iteration (p.72)

Check-in
1 $x = -13.768$ $x = -8.048$
2 $x = -0.4$ (range accepted $-0.5 \leq x \leq -0.3$)
$x = -5.6$ (range accepted $-5.7 \leq x \leq -5.5$)

Set?
A $x = -4$ gives -17 and $x = -3$ gives 14 and since 0 is between -17 and 14, x must be between -4 and -3
B $x = 1.6$ gives -0.904 and $x = 1.7$ gives 0.413 and since 0 is between -0.904 and 0.413, x must be between 1.6 and 1.7
C $x = 4$ gives -7 and $x = 5$ gives 18 and since 0 is between -7 and 18, x must be between 4 and 5
D $x_5 = -0.73$ (to 2 decimal places)
E $x_3 = 4.374$ (to 3 decimal places)

Go!
1 a) $x = 1$ gives 3 and $x = 2$ gives -8 and since 0 is between -8 and 3, x must be between 1 and 2
b) $x^2 - 2x^3 = -4$
$x^2(1 - 2x) = -4$
$x^2 = \frac{-4}{1 - 2x}$
$x^2 = \frac{4}{2x - 1}$
$x = \sqrt{\frac{4}{2x - 1}}$
c) $x_4 = 1.4670(67808)$

2 Between $x = 1.1$ and $x = 1.2$

3 2028

4 a) $x_1 = 1.27(272727)$
$x_2 = 1.6241(61074)$
$x_3 = 1.4525(98717)$
$x_4 = 1.5367(65518)$
b) Continuing
$x_5 = 1.4954(63151)$
$x_6 = 1.5157(40417)$
$x_7 = 1.5057(86143)$
$x_8 = 1.5106(73137)$
so $x = 1.51$ to 2 decimal places.

Inequalities 1 (p.74)

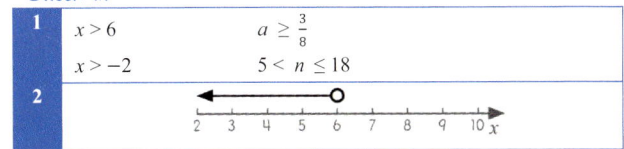

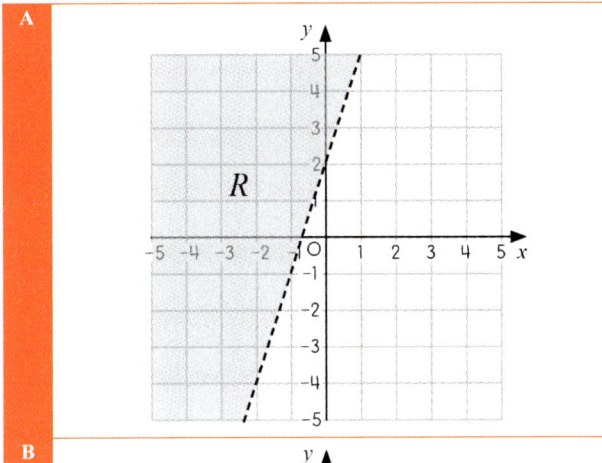

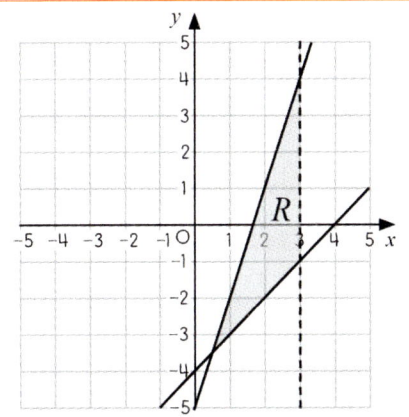

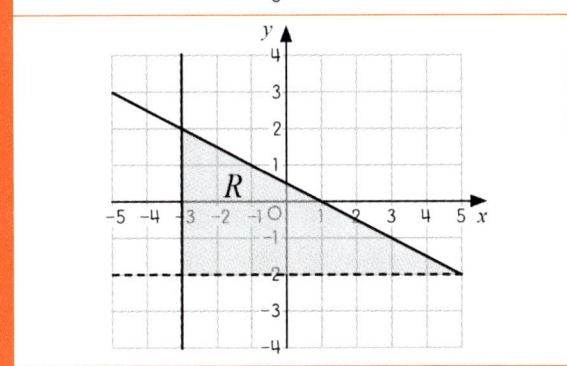

Check-in
1 $x > 6$ $a \geq \frac{3}{8}$
$x > -2$ $5 < n \leq 18$
2 [number line showing open circle at 6 with arrow extending left, marks from 2 to 10]

Set?
A [graph showing region R to the left of a dashed steep line]
B [graph showing region R bounded by two solid lines and a dashed vertical line]
C [graph showing region R bounded by solid lines and a dashed horizontal line at y = -2]

Go!
1 [graph showing region R bounded by a solid line, a dashed line, and a vertical line]

2 Lucus should have included points on the solid line and excluded points on the dotted line. There are only three points: $(-2, 1), (-1, 1)$ and $(0, 1)$

3 $y \leq \frac{1}{4}x + 3$
$y \leq -2x + 4$
$y > -x + 1$ or equivalents

4 $(0, 3), (0, 2), (0, 1), (0, 0), (0, -1), (1, 2)$ and $(1, 1)$

143

Inequalities 2 (p.76)

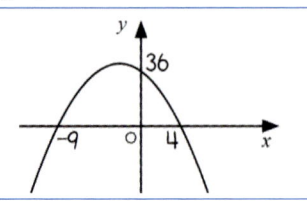

Check-in

1	$x = -7$ and $x = 1$
2	$x = -3.5$ and $x = 5$
3	(graph of downward parabola with roots at -9 and 4, y-intercept 36)

Set?

A	$-7 < x < 6$
B	$x \leq -5$ and $x \geq -1$
C	$6 \leq x \leq 8$
D	$x \leq 2.5$ and $x \geq 8$
E	$x < -0.6$ and $x > 2$
F	$x \leq -8$ and $x \geq 2$
G	$x < -2$ and $x > 4$
H	$-\frac{5}{3} \leq x \leq -\frac{3}{2}$

Go!

1	Muriel has not considered both solutions to $g^2 < \frac{25}{4}$ (positive and negative).
2	$-6, -5, -4, -3, -2, -1, 0, 1, 2$
3	$x < -1.60$ and $x > 0.94$
4	$-5, -4, -3, -2, -1, 0, 1, 2$
5	$a = 2, b = -19$ and $c = 24$

Sequences 1 (p.78)

Check-in

1	$-13, -5, 3, 11, 19, 27, 35$ $20, 14, 8, 2, -4, -10, -16$ $8, 9, 17, 26, 43, 69, 112$
2	$2, 7, 12, 17$
3	76

Set?

A	Ticks against the 2nd, 5th, 6th and 8th sequences.
B	$20\sqrt{2}, 40$
C	$-11,764.9$
D	$\frac{27}{128}$ (or $-\frac{27}{128}$)
E	14.5
F	20

Go!

1	$\frac{160}{3}$ or equivalent
2	$7 + 2\sqrt{3}$
3	a) $\frac{9}{10}$ or equivalent b) A 10% decrease in population each year.
4	a) $k = 8$ b) $r = 1 + \sqrt{2}$ so the fifth term is $(\sqrt{8} + 2)(1 + \sqrt{2})(1 + \sqrt{2})$ $= (4\sqrt{2} + 6)(1 + \sqrt{2})$ $= 14 + 10\sqrt{2} = 2(7 + 5\sqrt{2})$

Sequences 2 (p.80)

Check-in

1	$4n + 23$ $2n - 11$ $-3n + 13$ or $13 - 3n$ $4.5n + 9$

Set?

A	Ticks against 1st and 3rd sequences.
B	734
C	(i) $5n^2 + 2n$ (ii) $2n^2 + 7n - 2$ (iii) $3n^2 + n + 5$
D	(i) $n^2 - 2n + 4$ (ii) $2n^2 - 5n + 10$

Go!

1	a) $n^2 - 6n + 5$ b) $\frac{1}{4}n^2 - \frac{1}{2}n + 3$
2	$2n^2 - 5n + 7$
3	The nth term of $3, 7, 15, 27, 43 \ldots$ is $2n^2 - 2n + 3$
4	1st differences $= a + b, a + 3b, a + 5b, a + 7b$ giving a 2nd difference of $2b$ so the nth term starts bn^2 the adjustment $= 5a - b, 6a - 3b, 7a - 5b, 8a - 7b, 9a - 9b$ which has an nth term of $(a - 2b)n + 4a + b$ so $bn^2 + (a - 2b)n + 4a + b$ $= bn^2 + an - 2bn + 4a + b$ $= b(n^2 - 2n + 1) + a(n + 4)$ $= b(n - 1)^2 + a(n + 4)$

Direct Proportion (p.82)

Check-in

1.

x	5	8
y	35	56

x	6	10
y	9	15

x	12	7
y	3	1.75

x	12	15
y	49	35

2	Graph B

Set?

A	(i) $y = 3x$ (ii) $y = 0.25x$ (iii) $y = 5.2x$ or equivalents
B	270
C	115
D	34.56
E	7
F	3.61

Go!

1	No, y is directly proportional to x^2 not x. When $x = 5$ the correct value of y is 1000
2	44.1 metres per second
3	$y = kx^2$ and $x = mw$ $y = k(mw)^2 = km^2w^2 = rw^2$ (k, m, km^2 and r are multipliers) so $9072 = r \times 9^2 = 81r$ and $r = 112$ so $y = 112w^2$ and when $w = 8$, $y = 7168$
4	$T = k\sqrt{L}$ A reduction of 19% uses a multiplier of 0.81 So $T = k\sqrt{0.81 \times L} = k \times \sqrt{0.81} \times \sqrt{L} = 0.9k\sqrt{L}$ The multiplier 0.9 represents a 10% reduction in T.

Inverse Proportion (p.84)

Check-in

1.

x	3	12
y	40	10

x	25	20
y	24	30

x	50	7.5
y	3	20

x	15	20
y	6.4	4.8

2	Graph A

Set?

A	(i) $y = \frac{60}{x}$ (ii) $\frac{3}{2}$ (or 1.5)
B	(i) $p = \frac{288}{q}$ (ii) 48
C	$\frac{1}{8}$ (or 0.125)
D	100
E	$\frac{3}{5}$ (or 0.6)
F	$\frac{16}{25}$ (or 0.64)

Go!

1 a) Year 131
b) For example: The workers all worked at the same rate.

2 50% reduction.

3 $a = kb^2$ and $b = \frac{m}{c}$

so $a = k\left(\frac{m}{c}\right)^2 = \frac{km^2}{c^2} = \frac{r}{c^2}$ (k, m, km^2 and r are multipliers)

so $1.8 = \frac{r}{10^2}$ leading to $r = 180$ so $a = \frac{180}{c^2}$

When $a = \frac{5}{4}$, $c^2 = 144$ and the positive value of $c = 12$

4 $y = \frac{k}{\sqrt{x}}$ and $x = mw^3$

so $y = \frac{k}{\sqrt{mw^3}} = \frac{k}{\sqrt{m}\sqrt{w^3}} = \frac{r}{\sqrt{w^3}}$ (k, m, $\frac{k}{\sqrt{m}}$ and r are multipliers)

so $0.375 = \frac{r}{\sqrt{4^3}}$ leading to $r = 3$ so $y = \frac{3}{\sqrt{w^3}}$

When $w = 9$, $y = \frac{1}{9}$

Enlargements (p.86)

Check-in

1 Enlargement, scale factor of 2, centre of enlargement (4, 4)

2 [graph]

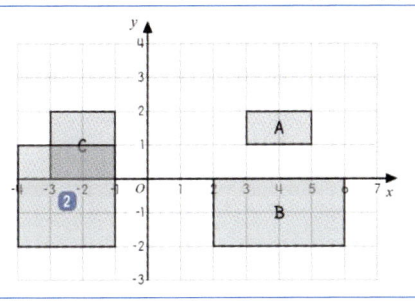

Set?

A, B, C [graph]

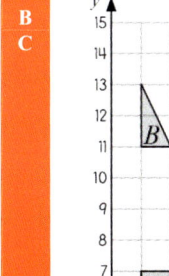

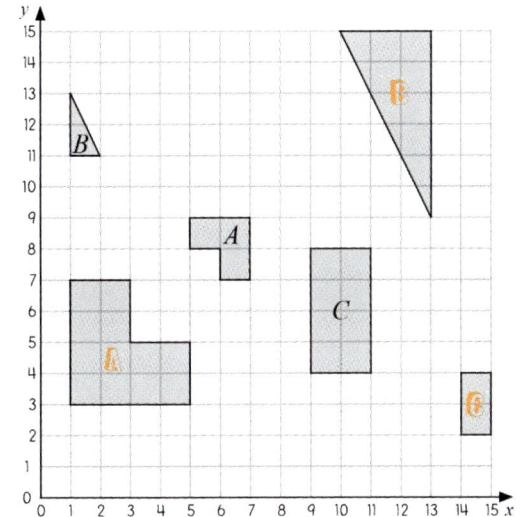

D Enlargement, scale factor of –1.5, centre of enlargement (3, 1)

Go!

1 No, she had not joined the matching vertices.
The centre of enlargement is (–1, 0) and the scale factor is –2

2 [graph showing P and Q]

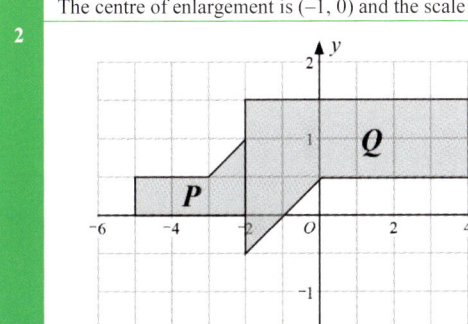

3 $-\frac{10}{7}$

4 $(-37, -37)$

Transformations (p.88)

Check-in

1 [graph]

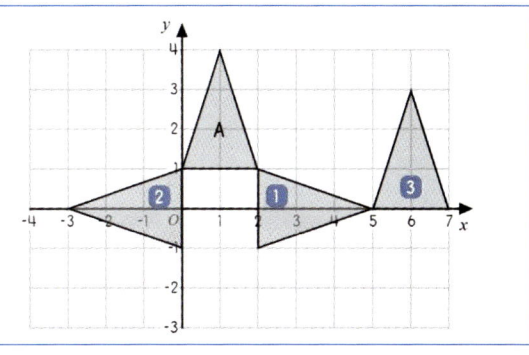

Set?

[graph]

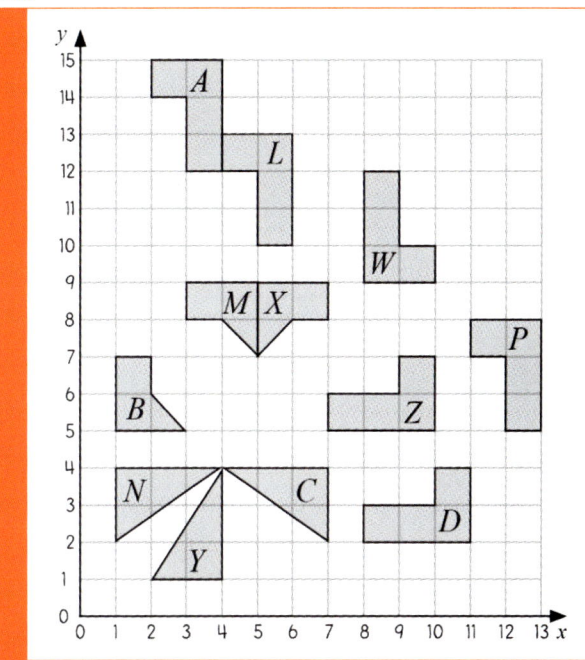

A	Rotation, 180°, centre of rotation (6, 12)
B	Rotation, 90° clockwise (or 270° anticlockwise), centre of rotation (5, 5)
C	Rotation, 90° clockwise (or 270° anticlockwise), centre of rotation (4, 4)
D	Translation using column vector $\begin{pmatrix}-1\\3\end{pmatrix}$
E	M to X at (5, 7) and (5, 9) C to N at (4, 4) N to Y at (4, 4) C to Y at (4, 4)

Go!

1 $a = -2$ and $b = -2$

2 a) 0
b) 1
c) 0

3 There is only one invariant point at (2, 1.25)

4 Sometimes true.
There is one invariant point if the centre of enlargement is on the original shape.
There are no invariant points in all other cases.

145

Circle Theorems 1 (p.90)

Check-in

1 Diagram showing a circle with labels: circumference, radius, diameter, chord.

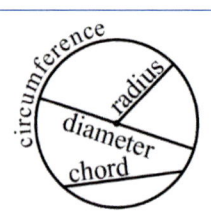

2 Two segments.

Set?

A
(i) $x = 57$
Angles in the same segment are equal.
(ii) $y = 66$
The angle at the centre is double the angle at the circumference.
(iii) $a = 12$
The angle in a semicircle is a right angle.
(iv) $b = 40$
Angles in the same segment are equal, angles on a straight line add up to 180° and angles in a triangle add up to 180°
(v) $p = 56$
Angles in the same segment are equal and angles in a triangle add up to 180°
(vi) $q = 72$
Angle $OBA = 18°$ Base angles in an isosceles triangle are equal.
Angle $AOB = 144°$ Angles in a triangle add up to 180° and the angle at the centre is double the angle at the circumference.

B 127°
Angles in the same segment are equal (angle $JMN = 21°$) and angles in a triangle add up to 180°

Go!

1 No. They are not in the same segment.

2 $x = 38$
Angle $BOC = 104°$ Angles on a straight line add up to 180°
Angle $OBC = 38°$ Base angles in an isosceles triangle are equal and angles in a triangle add up to 180°

$x = 38$
Angle $ABC = 38°$ The angle at the centre is double the angle at the circumference and base angles in an isosceles triangle are equal.

3 $a = 62°$
Circle theorem used: The angle at the centre is double the angle at the circumference.

4

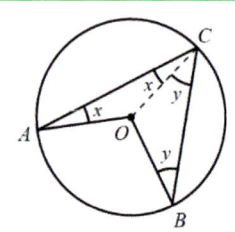

Let: angle $OAC =$ angle $OCA = x$ (Isosceles triangle)
Let: angle $OBC =$ angle $OCB = y$ (Isosceles triangle)
Angle $AOC = 180 - 2x$ and angle $BOC = 180 - 2y$
so angle $AOB = 360 - (180 - 2x) - (180 - 2y) = 2x + 2y$
Angle $ACB = x + y$
so angle $AOB = 2 \times$ angle ACB

Circle Theorems 2 (p.92)

Check-in

1 A **tangent** is a straight line outside a circle that touches the circle exactly once.
A **chord** (or **diameter**) splits a circle into two segments.

2 $AC = 17$ cm $AC = 55$ mm

Set?

A
(i) $x = 64$
Base angles in an isosceles triangle are equal and a radius and tangent that meet are perpendicular.
(ii) $a = 53$
Base angles in an isosceles triangle are equal and a radius and tangent that meet are perpendicular.
(iii) $y = 6$
A radius that is perpendicular to a chord also bisects the chord.
(iv) $p = 14$
A radius that is perpendicular to a chord also bisects the chord (so forms a right angled triangle) and angles in a triangle add up to 180°
(v) $b = 19$
Two tangents that meet at a point are equal in length.
(vi) $q = 144$
A radius and tangent that meet are perpendicular, base angles in an isosceles triangle are equal and angles in a triangle add up to 180°

B
(i) $v = 8$ (cm)
(ii) $t = 25$ (cm)

C 7.8 cm

Go!

1 No. The line through Q and P is not a tangent.

2 $y = 70$
A radius and tangent that meet are perpendicular, the angles in the quadrilateral $BDAO$ add up to 360° and the angle at the centre is double the angle at the circumference.

3 $x = 62$
Possible reasons:
Angle $ADB = 28°$ Corresponding angles are equal.
Angle $ADC = 90°$ The angle in a semicircle is a right angle.
So $x = 90 - 28 = 62$

4

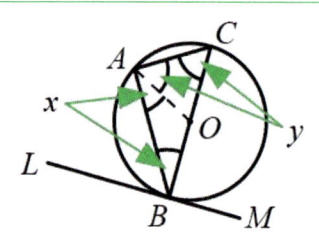

Possible proof:
Let: angle $BAO =$ angle $ABO = x$
Let: angle $OAC =$ angle $OCA = y$ (this is angle ACB)
From triangle ABC we get $2x + 2y = 180°$ so $x + y = 90°$
Angle $LBO = 90°$ (A radius and tangent that meet are perpendicular) so angle $LBO = x + y$
and angle $ABO = x$ so angle ABL must equal y
so angle $ABL =$ angle ACB

Circle Theorems 3 (p.94)

Check-in

1 $x = 63°$ $x = 54°$
$x = 90°$ $x = 41°$

Set?

A
(i) $a = 76$
Opposite angles in a cyclic quadrilateral add up to 180°
(ii) $s = 61$
The alternate segment theorem.
(iii) $x = 68$ and $y = 107$
Opposite angles in a cyclic quadrilateral add up to 180°
(iv) $w = 68$
The alternate segment theorem.
$v = 21$
Angles in a triangle add up to 180°
(v) $b = 34$
The alternate segment theorem and opposite angles in a cyclic quadrilateral add up to 180°
(vi) $h = 87$
The alternate segment theorem and opposite angles in a cyclic quadrilateral add up to 180°

B
(i) 57°
(ii) 52°
(iii) 99°

146

Go!

1	The 61° he has labelled is between the line AC and the circumference of the circle. He should have labelled between lines AC and AQ.
2	101°
3	Triangle OQP is isosceles so angle OPQ = 62° $x = 28°$ because the angle between a tangent and a radius that meet is 90° Angle $y = 180 – (26 + 62) = 92°$ because opposite angles in a cyclic quadrilateral add up to 180° $28° + 92° = 120°$
4	Triangles AOD, DOC, COB and BOA are isosceles. $2w + 2x + 2y + 2z = 360°$ so $w + x + y + z = 180°$ And from the diagram $a + c = 180°$ and $b + d = 180°$

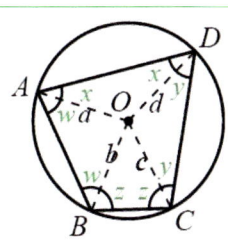

Equation of a Circle 1 (p.96)

Check-in

1	$x = 17$ (cm) $x = 3$ (cm) $x = 7$ (mm) $x = 15$ (cm)

Set?

A	(i) 3 (ii) 10 (iii) 3.5 (iv) 1.5 (or $\frac{3}{2}$)
B	$x^2 + y^2 = 2.25$
C	$n = -4.47$
D	

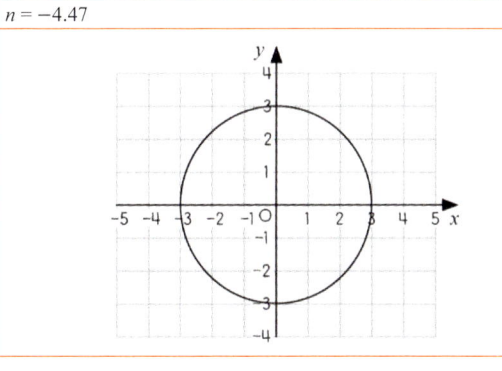

Go!

1	Jeevan has not used the square of the radius (7^2). The correct answer is $x^2 + y^2 = 49$
2	a) $x = 1.9, y = 8.8$ and $x = -5.9, y = -6.8$ b)
3	a) $a = 8$ b) (15, 8) c) (11.5, 11.5)
4	(−5, 12)

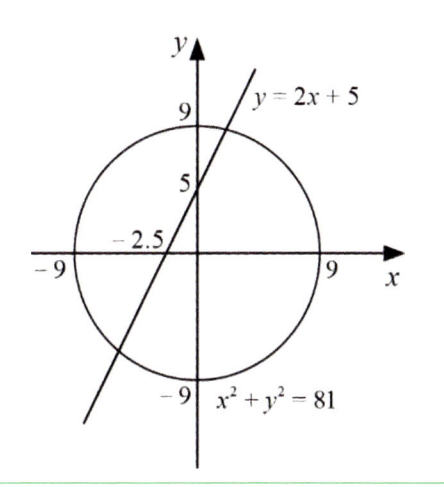

Equation of a Circle 2 (p.98)

Check-in

1	5 −2 0.25 −2
2	$-\frac{5}{2}$ (or −2.5)
3	$y = 5x - 2$

Set?

A	$y = \frac{3}{2}x - 13$ or equivalent
B	$y = \frac{2}{5}x + \frac{29}{5}$ or equivalent
C	$y = -\frac{1}{2}x + 10$ and $y = \frac{1}{2}x - 10$ or equivalents

Go!

1	$y = -\frac{5}{4}x + \frac{41}{4}$ or equivalent
2	(10, 0)
3	10.2 units^2
4	$x^2 + y^2 = 40$

Similarity (p.100)

Check-in

1	7.5 (cm)
2	$m = 15$ $n = 8.4$

Set?

A	27 : 64
B	5 : 2
C	5 : 4
D	9 : 4
E	27 : 64
F	250 (cm^2)
G	166 (times greater)

Go!

1	2.4 (m^2) Length scale factor = 2 so volume scale factor = $2^3 = 8$ The contents (volume) of the tin covers 0.3 m^2 so $0.3 \times 8 = 2.4$
2	a) $\frac{1}{64}$ b) For example: Assume that the earth and the moon are perfect spheres.
3	Yes, she is correct. The length scale factor is 4 so the volume scale factor is 64 and $2368 ÷ 64 = 37$
4	159 (cm^2)

Solids (p.102)

Check-in

1	Surface area = 360 cm^2 Volume = 300 cm^3
2	24π cm^2
3	164.9 cm^2
4	80 cm^3
5	288π cm^3

Set?

A	324π cm^3
B	204 cm^3
C	723 cm^2
D	$4n^3 + \pi n^3$ or equivalent

Go!

1	1.7 m
2	3079 grams
3	8.98 cm
4	a) 7248π cm^3 b) 600π cm^2

147

Trigonometry in 3D (p.104)

Check-in
1	8.5 (cm)	9.3 (cm)
	3.8 (cm)	54.0 (°)

Set?
A	(i) 10.4 cm
	(ii) 90°
	(iii) 11.2 cm
B	(i) 7.8 cm
	(ii) 23.6 cm
	(iii) 18.3°
C	(i) 14.1 cm
	(ii) 59.5° (accept 59.6° if 14.1 cm was used)

Go!
1	65.9°
2	27.2°
3	221 cm²
4	12.3°

The Sine Rule (p.106)

Check-in
1	10.5 (cm)	7.8 (cm)
	43.8 (°)	10.2 (cm)

Set?
A	(i) $x = 4.12$ (cm)
	(ii) $y = 8.32$ (cm)
	(iii) $z = 15.4$ (cm)
B	(i) 64.5 (°)
	(ii) 56.2 (°)
	(iii) 34.3 (°)
C	41.8°

Go!
1	Yes, Hattie is correct. The sine rule can also be written as $\frac{\sin A}{a} = \frac{\sin B}{b} = \frac{\sin C}{c}$ which is useful for finding angles.
2	42.7°
3	6.208 (cm)
4	176°
5	133.6° You may have an answer of 46.4° which would be correct for an acute angle at *LMN*, however angle *LMN* is an obtuse angle so we subtract our answer from 180°. This is known as the ambiguous case of the sine rule.

The Cosine Rule (p.108)

Check-in
1	6.9 (cm)	86.9 (°)
2	*Triangle with A, B=7.6 cm, 8.3 cm, angle 58° at C*	

Set?
A	(i) $p = 13.9$ (cm)
	(ii) $q = 9.22$ (cm)
	(iii) $r = 18.8$ (cm)
B	(i) 33.6 (°)
	(ii) 75.9 (°)
	(iii) 85.4 (°)
C	13 m

Go!
1	Fintan has not used the correct order of operations, he has worked out 36 + 81 − 108 first. The correct answer is 118.8°
2	$x = 24.6$ (cm)
3	$A = 76.4°$ $B = 45.6°$ $C = 58.0°$
4	130°

Area of a Triangle (p.110)

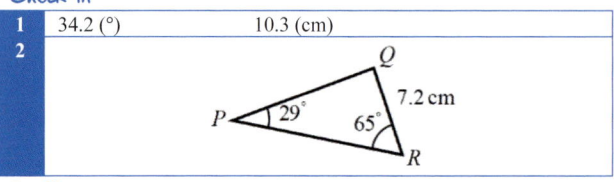

Check-in
1	34.2 (°)	10.3 (cm)
2	*Triangle PQR, angle P = 29°, angle R = 65°, QR = 7.2 cm*	

Set?
A	(i) 128 m²
	(ii) 43.1 cm²
	(iii) 323 cm²
B	(i) 13.4 (cm)
	(ii) 84.0 (°)
	(iii) 5.8 (cm)
C	28 cm²

Go!
1	60°
2	110.5 cm²
3	53 cm²
4	Yes, because sin 90° = 1 so is equivalent to $\frac{1}{2} \times$ base $\times$ height

Geometric Proof (p.112)

Check-in
1	Ticks against 1st, 3rd, 4th and 5th statements.
2	135°
3	1.5 (cm)

Set?
A	For example: Angle *BXC* = angle *AXE* as vertically opposite angles are equal. Angle *ACB* = angle *AEB* as triangles *ABC* and *ABE* are isosceles so angle *CBX* = angle *EAX* *AE* = *BC* as both are sides of a regular pentagon. Therefore triangles *BCX* and *AEX* are congruent (AAS).
B	For example: Angle *OAC* = angle *OBC* as both are right angles. *OC* is a shared side. *OA* = *OB* as they are both radii. Therefore triangles *ACO* and *BCO* are congruent (RHS) so *AC* = *BC*.
C	For example: Interior angles of a regular pentagon are 108° so angle *APQ* = angle *AQP* = 72° (Angles that meet at a point on a straight line add up to 180°). Angle *PAQ* = 36° (Angles in a triangle add up to 180°) also angle *CTS* = angle *BRS* = 72° and angle *SCT* = angle *SBR* = 72° (Triangle *ABC* is isosceles). So angle *BSR* = angle *CST* = 36° Therefore triangles *PAQ* and *CST* are similar (AAA).

Go!
1	Angle *AMC* = 102° as angles that meet at a point on a straight line add up to 180° so angle *AMC* = angle *ANB* Angle *BAC* is a common angle. Given *AB* = *AC* Therefore triangles *ABN* and *ACM* are congruent (AAS).
2	Triangle *ABC*: Using $\tan 60° = \frac{AB}{x}$ ($\tan 60° = \sqrt{3}$) So $AB = \sqrt{3}x$ Area of triangle $ABC = \frac{1}{2} \times x \times \sqrt{3}x = \frac{\sqrt{3}x^2}{2}$ Triangle *BCD*: Pythagoras' Theorem with triangle *ABC* we find $BC = 2x$ Using $\tan 60° = \frac{2x}{CD}$ so $CD = \frac{2x}{\sqrt{3}}$ ($\tan 60° = \sqrt{3}$) Area of triangle $BCD = \frac{1}{2} \times \frac{2x}{\sqrt{3}} \times 2x = \frac{2x^2}{\sqrt{3}}$ Total area $= \frac{\sqrt{3}x^2}{2} + \frac{2x^2}{\sqrt{3}} = \frac{(\sqrt{3} \times \sqrt{3}x^2) + (2 \times 2x^2)}{2\sqrt{3}} = \frac{3x^2 + 4x^2}{2\sqrt{3}} = \frac{7x^2}{2\sqrt{3}}$ Rationalising the denominator: $\frac{7 \times \sqrt{3}}{2\sqrt{3} \times \sqrt{3}} x^2 = \frac{7\sqrt{3}}{6} x^2$
3	Angle *BCD* = 90° (Co-interior angles add up 180°) so angle *DCE* = angle *ABE* *EA* = *ED* given that they are the same length. Let angle *BAE* = x so angle *BEA* = $180 − (90 + x) = 90 − x$ then angle *CED* = $180 − 90 − (90 − x) = x$ so angle *BAE* = angle *CED* Therefore triangles *ABE* and *ECD* are congruent (AAS).

4	Area of $A = \frac{1}{2} \times a \times a \times \sin 60° = \frac{\sqrt{3} a^2}{4}$ $\left(\sin 60° = \frac{\sqrt{3}}{2}\right)$
	Rearranging gives $a^2 = \frac{4 \times \text{Area of } A}{\sqrt{3}}$
	Area of $B = \frac{1}{2} \times b \times b \times \sin 60° = \frac{\sqrt{3} b^2}{4}$
	Rearranging gives $b^2 = \frac{4 \times \text{Area of } B}{\sqrt{3}}$
	Area of $C = \frac{1}{2} \times c \times c \times \sin 60° = \frac{\sqrt{3} c^2}{4}$
	Rearranging gives $c^2 = \frac{4 \times \text{Area of } C}{\sqrt{3}}$
	Substituting into Pythagoras' Theorem gives:
	$\frac{4 \times \text{Area of } A}{\sqrt{3}} + \frac{4 \times \text{Area of } B}{\sqrt{3}} = \frac{4 \times \text{Area of } C}{\sqrt{3}}$
	which simplifies to Area of A + Area of B = Area of C

Vectors 1 (p.114)

Check-in

1	$a = 3p$	$b = 4(p+q)$ or $4p + 4q$
	$c = -4(p+q)$ or $-4p - 4q$	$d = 2p + q$
2	$\frac{3}{5}a$	

Set?

A	(i) $-9a - b$
	(ii) $4(a - b)$ or $4a - 4b$
B	(i) $\overrightarrow{QR} = 2.5 \times \overrightarrow{PQ}$ so $\overrightarrow{PQ}$ and $\overrightarrow{QR}$ are parallel. Both vectors pass through Q so P, Q and R must lie on a straight line.
	(ii) $0.4 : 1$ or $\frac{2}{5} : 1$
C	$-\frac{3}{4}a - \frac{5}{8}b$
D	$\overrightarrow{AB} = a - b$ and $\overrightarrow{DE} = \overrightarrow{DC} + \overrightarrow{CE} = \frac{2}{5}b - \frac{2}{5}a = -\frac{2}{5}(a - b)$ so $\overrightarrow{AB}$ is parallel to $\overrightarrow{DE}$

Go!

1	$\overrightarrow{BD} = -b + a$ and $\overrightarrow{AM} = a - \frac{1}{2}(-b + a) = a + \frac{1}{2}b - \frac{1}{2}a = \frac{1}{2}a + \frac{1}{2}b$
	$\overrightarrow{AC} = a + b = 2 \times \overrightarrow{AM}$ so M is the midpoint of AC.
2	$\overrightarrow{ZP} = 2c$ and $\overrightarrow{WP} = d + 2c$
	$\overrightarrow{WN} = \frac{1}{2}(d + 2c)$ and $\overrightarrow{WM} = \frac{1}{2}(d + c) = \overrightarrow{MY}$
	$\overrightarrow{MN} = \overrightarrow{MW} + \overrightarrow{WN} = -\frac{1}{2}(d + c) + \frac{1}{2}(d + 2c) = \frac{1}{2}c$
	$\overrightarrow{ZP} = 4 \times \overrightarrow{MN}$ so $\overrightarrow{MN}$ is parallel to $\overrightarrow{ZP}$
3	$\overrightarrow{SY} = \frac{4}{5}a$ and $\overrightarrow{SX} = \frac{1}{3}b$
	$\overrightarrow{YX} = \overrightarrow{YS} + \overrightarrow{SX} = -\frac{4}{5}a + \frac{1}{3}b$ and $\overrightarrow{YP} = 2 \times \overrightarrow{YX} = -\frac{8}{5}a + \frac{2}{3}b$
	$\overrightarrow{SP} = \overrightarrow{SY} + \overrightarrow{YP} = \frac{4}{5}a + -\frac{8}{5}a + \frac{2}{3}b = -\frac{4}{5}a + \frac{2}{3}b = \frac{2}{15}(-6a + 5b)$
	So $k = \frac{2}{15}$
4	$\overrightarrow{AB} = 2a - 2c - 2b$ and $\overrightarrow{AL} = \frac{1}{2} \times \overrightarrow{AB} = a - c - b$ $(= \overrightarrow{LB})$
	$\overrightarrow{PL} = \overrightarrow{PA} + \overrightarrow{AL} = -a + (a - c - b) = -c - b$ and $\overrightarrow{NM} = -c - b$
	$\overrightarrow{LM} = \overrightarrow{LB} + \overrightarrow{BM} = (a - c - b) + b = a - c$ and $\overrightarrow{PN} = a - c$
	Two pairs of equal and parallel sides so $LMNP$ is a parallelogram.

Vectors 2 (p.116)

Check-in

1	$x = -7$
2	$k = 0.4$
3	$4a - 2b$ or equivalent

Set?

A	$\overrightarrow{AR} = 4a + 12b$ and $\overrightarrow{PB} = 8a + 4b$
	$\overrightarrow{PX} = \overrightarrow{PA} + \overrightarrow{AX} = 4a + m(4a + 12b) = (4 + 4m)a + 12mb$
	$\overrightarrow{PB}$ and $\overrightarrow{PX}$ are collinear so $\frac{1}{2}(4 + 4m) = 12m$
	and solving gives $m = \frac{1}{5}$
	Therefore $\overrightarrow{PX} = \frac{24}{5}a + \frac{12}{5}b = \frac{12}{5}(2a + b)$
	$\overrightarrow{AX} = \overrightarrow{AP} + \overrightarrow{PX} = -4a + \frac{12}{5}(2a + b) = \frac{4}{5}a + \frac{12}{5}b = \frac{4}{5}(a + 3b)$
	So $k = \frac{4}{5}$
B	$\overrightarrow{BD} = -2a + 6b$ so $\overrightarrow{BN} = -a + 3b$ and $\overrightarrow{MA} = -2b$
	$\overrightarrow{MN} = \overrightarrow{MA} + \overrightarrow{AB} + \overrightarrow{BN} = -2b + 2a + -a + 3b = a + b$
	$\overrightarrow{MP} = \overrightarrow{MD} + \overrightarrow{DP} = 4b + ka$
	Since $\overrightarrow{MN}$ and $\overrightarrow{MP}$ are collinear $k = 4$ so $\overrightarrow{MP} = 4b + 4a$
	and $\overrightarrow{MN} = a + b$ therefore $\overrightarrow{MN} = \frac{1}{4} \times \overrightarrow{MP}$

Go!

1	$\overrightarrow{YS} = -5a + 30b$ and $\overrightarrow{QR} = -10a + 15b$
	$\overrightarrow{YX} = \overrightarrow{YQ} + \overrightarrow{QX} = 5a + m(-10a + 15b) = (5 - 10m)a + 15mb$
	$\overrightarrow{YS}$ and $\overrightarrow{YX}$ are collinear so $-6(5 - 10m) = 15m$
	and solving gives $m = \frac{2}{3}$
	Therefore $\overrightarrow{YX} = -\frac{5}{3}a + 10b$
	$\overrightarrow{QX} = \overrightarrow{QY} + \overrightarrow{YX} = -5a + -\frac{5}{3}a + 10b = -\frac{20}{3}a + 10b = -\frac{10}{3}(2a - 3b)$
	So $k = -\frac{10}{3}$
2	$\overrightarrow{CF} = -2a$ and $\overrightarrow{BN} = 2b - \frac{3}{2}a$ so $\overrightarrow{BM} = b - \frac{3}{4}a$
	$\overrightarrow{CM} = \overrightarrow{CB} + \overrightarrow{BM} = -b + b - \frac{3}{4}a = -\frac{3}{4}a$
	So $m = -\frac{3}{4}$
3	$\overrightarrow{AD} = a + 2b$ and $\overrightarrow{PC} = 2a + b$
	$\overrightarrow{PQ} = \overrightarrow{PA} + \overrightarrow{AQ} = -b + m(a + 2b) = ma + (2m - 1)b$
	$\overrightarrow{PC}$ and $\overrightarrow{PQ}$ are collinear so $m = 2(2m - 1)$ and solving gives $m = \frac{2}{3}$
	Therefore $\overrightarrow{PQ} = \frac{2}{3}a + \frac{1}{3}b = \frac{1}{3}(2a + b)$ so $PQ : PC = \frac{1}{3} : 1 = 1 : 3$
	So $n = 3$
4	$\overrightarrow{AB} = 8a - 20b$ and $\overrightarrow{AX} = 2a - 5b$
	$\overrightarrow{XB} = 6a - 15b$ and $\overrightarrow{AY} = 6a$ and $\overrightarrow{YC} = 2a$
	$\overrightarrow{XY} = \overrightarrow{XA} + \overrightarrow{AY} = -2a + 5b + 6a = 4a + 5b$
	$\overrightarrow{YD} = \overrightarrow{YC} + \overrightarrow{CD} = 2a + m(20b)$
	$\overrightarrow{XY}$ and $\overrightarrow{YD}$ are collinear so $2 \times \frac{5}{4} = 20m$ and solving gives $m = \frac{1}{8}$
	Therefore $\overrightarrow{CD} = \frac{1}{8} \times 20b = \frac{5}{2}b$
	So $k = \frac{5}{2}$

Probability 1 (p.118)

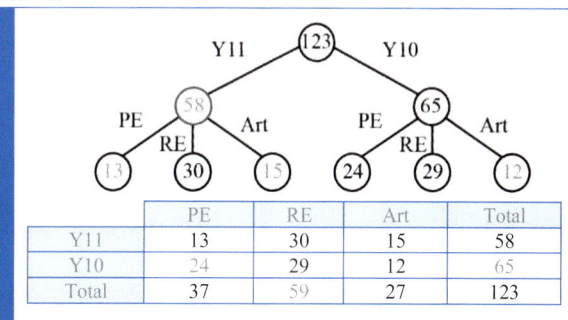

Check-in

1	

	PE	RE	Art	Total
Y11	13	30	15	58
Y10	24	29	12	65
Total	37	59	27	123

Set?

A	$\frac{14}{110}$ or equivalent
B	(i) $\frac{23}{51}$
	(ii) $\frac{10}{80}$ or equivalents
C	(i)

	Male	Female	Total
Finance	14	23	37
Sales	23	12	35
Ops	6	9	15
Total	43	44	87

(ii) $\frac{6}{43}$ or equivalent

Go!

1	$\frac{7}{28}$ or equivalent
2	$\frac{3}{53}$ or equivalent
3	$\frac{4}{36}$ or equivalent
4	Let x = number of students that study both PE and history, so we know: $60 + 48 - x + N = 240$
	And solving gives: $x = N - 132$
	So P(study PE given that they study history) = $\frac{x}{60} = \frac{N - 132}{60}$

149

Probability 2 (p.120)

Check-in

1

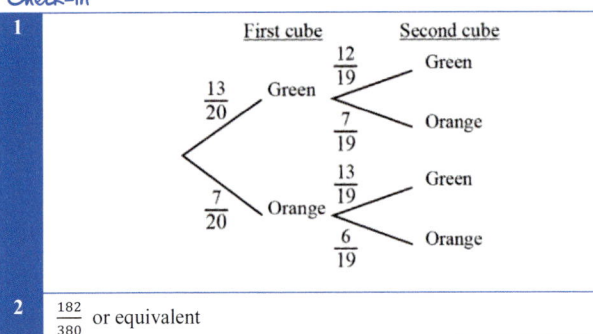

First cube → Second cube

- $\frac{13}{20}$ Green → $\frac{12}{19}$ Green
- Green → $\frac{7}{19}$ Orange
- $\frac{7}{20}$ Orange → $\frac{13}{19}$ Green
- Orange → $\frac{6}{19}$ Orange

2 $\frac{182}{380}$ or equivalent

Set?

A (i) Given the number of green counters is $2n$ and the ratio Red : Green = 3 : 2 the total counters can be written as $5n$.

P(Green) as the first counter $= \frac{2n}{5n}$ and

P(Green) as the second counter $= \frac{2n-1}{5n-1}$

P(Green, Green) $= \frac{2n}{5n} \times \frac{2n-1}{5n-1} = \frac{1}{7}$

$= \frac{2n(2n-1)}{5n(5n-1)} = \frac{1}{7}$

$= 7 \times 2n(2n-1) = 5n(5n-1)$

$= 14n(2n-1) = 5n(5n-1)$

Dividing both sides by n gives $14(2n-1) = 5(5n-1)$

(ii) 6 (counters)

B P(Red, Green) $= \frac{5}{n+5} \times \frac{n}{n+4} = \frac{5n}{(n+5)(n+4)}$ and

P(Green, Red) $= \frac{n}{n+5} \times \frac{5}{n+4} = \frac{5n}{(n+5)(n+4)}$

P(Red, Green) + P(Green, Red) $= \frac{5n}{(n+5)(n+4)} + \frac{5n}{(n+5)(n+4)} = \frac{10}{21}$

$= \frac{10n}{(n+5)(n+4)} = \frac{10}{21}$

$= 21 \times 10n = 10(n+5)(n+4)$

$= 210n = 10(n^2 + 9n + 20)$

Dividing both sides by 10 gives $21n = n^2 + 9n + 20$

Rearranging gives $0 = n^2 - 12n + 20$

C P(Not red, Not red, Not red) $= \frac{n-6}{n} \times \frac{n-7}{n-1} \times \frac{n-8}{n-2}$

$= \frac{(n-6)(n-7)(n-8)}{n(n-1)(n-2)}$

$= \frac{n^3 - 21n^2 + 146n - 336}{n^3 - 3n^2 + 2n}$

Go!

1 a) P(Blue, Blue) $= \frac{x-2}{x} \times \frac{x-2}{x}$

$= \frac{x^2 - 4x + 4}{x^2}$

b) Dom needs to solve: $\frac{x^2 - 4x + 4}{x^2} = \frac{64}{81}$

$= 81(x^2 - 4x + 4) = 64x^2$

$= 81x^2 - 324x + 324 = 64x^2$

$= 17x^2 - 324x + 324 = 0$

Solving gives $x = 18$ (the only integer solution)

2 P(Black) $= \frac{2n}{5n}$

Adding 4 more black beads gives P(Black) $= \frac{2n+4}{5n+4} = \frac{1}{2}$

So $2(2n + 4) = 5n + 4$ and solving this gives $n = 4$

Therefore $2 \times 4 = 8$ black beads and $5 \times 4 = 20$ beads in total

$20 - 8 = 12$ blue beads

3 P(Yellow) $= \frac{5n}{8n}$

P(Yellow, Yellow) $= \frac{5n}{8n} \times \frac{5n-1}{8n-1} = \frac{35}{92}$

$= \frac{25n^2 - 5n}{64n^2 - 8n} = \frac{35}{92}$

$= 2300n^2 - 460n = 2240n^2 - 280n$

Rearranging gives $60n^2 - 180n = 0$

Solving gives $n = 3$ (the only valid solution)

The number of green counters is $3n = 3 \times 3 = 9$

4 P(Red, Red) $= \frac{n}{6+n} \times \frac{n-1}{6+n-1} = \frac{2}{15}$

$= \frac{n^2 - n}{n^2 + 11n + 30} = \frac{2}{15}$

$= 15n^2 - 15n = 2n^2 + 22n + 60$

Rearranging gives $13n^2 - 37n - 60 = 0$

Solving gives $n = 4$ (the only valid solution)

Total = 10 (6 black and 4 red)

Statistics (p.122)

Check-in

1
113
24.8
57

Set?

A (i) LQ = 105 and UQ = 121
(ii) LQ = 21 and UQ = 30.3
(iii) LQ = 28 and UQ = 40
(iv) LQ = 23 and UQ = 28.5

B (i) IQR = 16
(ii) IQR = 9.3
(iii) IQR = 12
(iv) IQR = 5.5

C For example:
On average, Class A are taller than Class B as the median is greater.

Class A has more consistent heights as they have a smaller interquartile range (or smaller range).

Go!

1 No. We do not compare highest/lowest values or quartiles, we compare median and IQR (or range).

2 Zain has worked out the range, not the IQR.
The correct answer is $70 - 51 = 19$

3 a) $60 \leq t < 90$
b) $60 \leq t < 90$
c) We do not know the exact values for the LQ and UQ.

4

	Group B	Group A
LQ	155	156
Median	164	164
UQ	171	171
IQR	16	15
Range	33	35

For example:
On average, Group A and Group B have similar armspans as they have the same median.

Group A have more consistent armspans as the interquartile range is smaller.

(Note: Group A have a larger range so their armspans are more spread out).

Box Plots (p.124)

Check-in

1. Median = 63, Interquartile range = 15

Set?

A. [Box plot showing data in kilograms, range approximately 12 to 57, box from ~22 to ~37, median ~30]

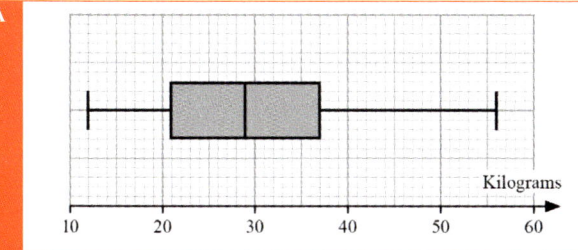

B. [Two box plots showing Number of goals for Team 1 and Team 2]

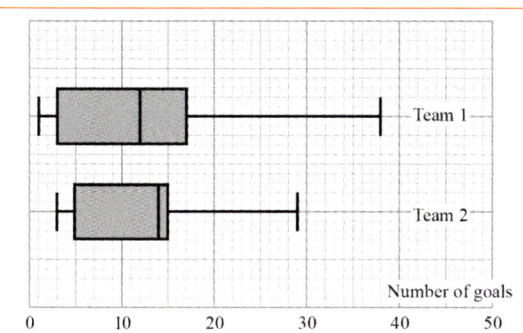

For example:
On average, Team 2 scored more goals than Team 1 as the median is greater.
The number of goals scored by Team 2 were more consistent as the interquartile range (or range) is smaller.

Go!

1. 15 numbers such that:
   Lowest = 124,
   LQ = 128, Median = 134, UQ = 146 and
   Greatest = 160

2. The median for Newtown is 40 so half of the people in Newtown are over 40 but for Oldtown half of the people are over 36 so less than half will be over 40.

3. For example:
   On average, the number of visitors increases over time as the median is getting larger.
   The spread is increasing over time, (as the range and/or IQR gets larger) which means that the number of visitors is more variable.

4. [Box plot on cm scale from 40 to 90, box approximately 52 to 72, median ~60]

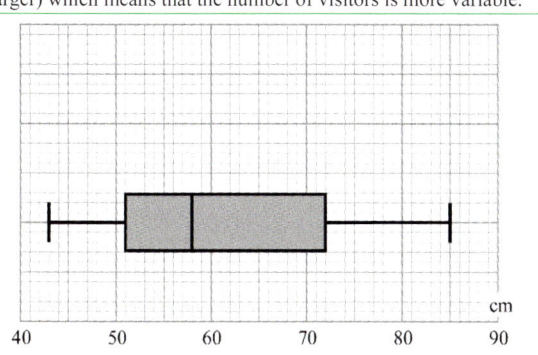

Cumulative Frequency (p.126)

Check-in

1. For example:
   On average, Aishah scored more points than Mia as her median is greater.
   Mia's scores are more consistent as her interquartile range (or range) is smaller.

Set?

A. [Cumulative frequency curve against Height (h cm), from 10 to 20]

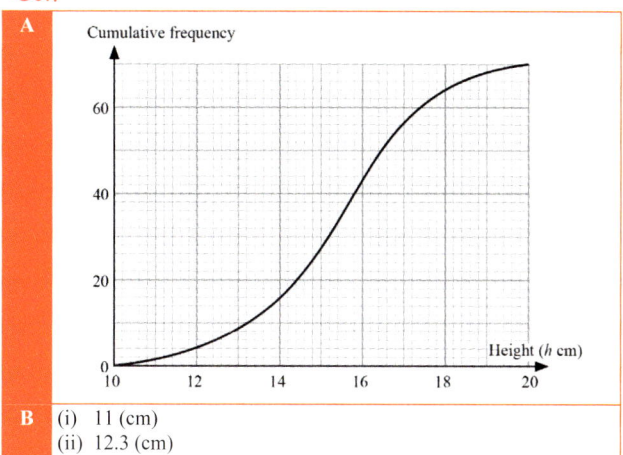

B. (i) 11 (cm)
   (ii) 12.3 (cm)
   (iii) 44 (worms)

Go!

1. Lois has not plotted the upper end of each interval.
   The height of the third point should be 44 not 48.

2.
	Men	Women
LQ	34.5	40
Median	39	45
UQ	45.5	57
IQR	11	17

For example:
On average, the men were faster than the women as the median is lower.
The men had more consistent times as the interquartile range is smaller.

3. $\dfrac{104}{120}$

4. [Box plot and cumulative frequency curve against Height (cm), from 140 to 190]

151

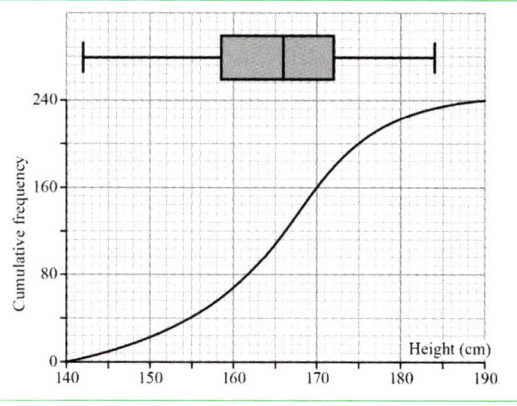

Histograms (p.128)

Check-in

1	$20 \leq m < 30$
2	$30 \leq m < 40$

Set?

A [Histogram: Frequency density vs Area (A m²), with bars from 50–150]

B (i) 44 (schools)
   (ii) 13 (schools)

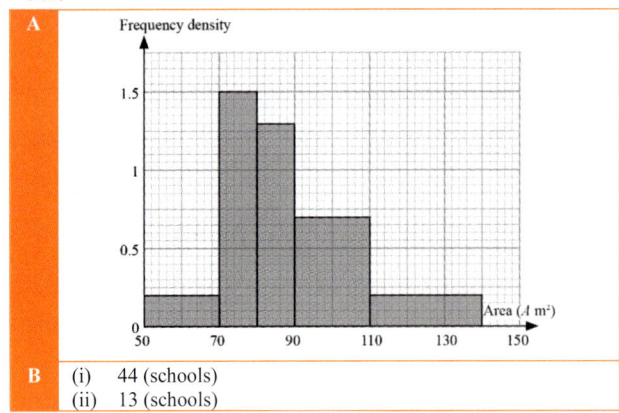

Go!

1 Tick against the final chart.
Only the final chart is not a histogram as it has frequency and unequal bars.
Note: The second chart is a histogram because the area of the bars are proportional to their frequencies.

2 a) Total frequency = 480 so LQ is the 120th value
$100 \leq$ Rainfall < 250, frequency is $0.6 \times 150 = 90$
$250 \leq$ Rainfall < 300, frequency is $3 \times 50 = 150$
120th value is $90 + \frac{1}{5}$ of 150
So LQ = 260 (mm)
b) $375 - 260 = 115$ (mm)

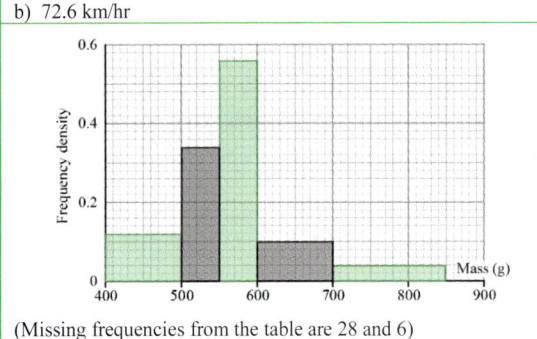

3 a) 72 km/hr
b) 72.6 km/hr

4 [Histogram: Frequency density vs Mass (g), bars from 400–900]

(Missing frequencies from the table are 28 and 6)

Capture-Recapture (p.130)

Check-in

1	$a = 375$	$b = 16$
	$c = 175$	$d = 9$

Set?

A (i) 200
   (ii) For example:
   No marks have rubbed off.
   No butterflies have died.

B (i) 96
   (ii) For example:
   No snails have died.
   None of the numbers have rubbed off.

C (i) 571 (to the nearest whole number)
   (ii) For example:
   No marks have come off.

Go!

1 a) 960
b) For example:
No marks have rubbed off.
No bees have died.

2 Yes. The calculation is one method to solve $\frac{100}{N} = \frac{12}{500}$ because $\frac{100}{12}$ is equivalent to $\frac{25}{3}$

3 Although Caz caught the most fish and would give the most accurate individual estimate, combining the results would give us a larger sample size and so the most accurate estimate.
$\frac{16}{60} = \frac{100}{N}$ so N = 375

4 $F = 128$ (fish)